Conference Proceedings of the Society for Experimental Mechanics Series

Series Editor

Kristin B. Zimmerman, Ph.D.
Society for Experimental Mechanics
Bethel, CT, USA

More information about this series at http://www.springer.com/series/8922

Barton C. Prorok • LaVern Starman
Editors

MEMS and Nanotechnology, Volume 5

Proceedings of the 2015 Annual Conference on Experimental and Applied Mechanics

Springer

Editors
Barton C. Prorok
Auburn University
Auburn, USA

LaVern Starman
Air Force Research Laboratory
Wright-Patterson AFB, OH, USA

ISSN 2191-5644 ISSN 2191-5652 (electronic)
Conference Proceedings of the Society for Experimental Mechanics Series
ISBN 978-3-319-22457-2 ISBN 978-3-319-22458-9 (eBook)
DOI 10.1007/978-3-319-22458-9

Library of Congress Control Number: 2015951228

Springer Cham Heidelberg New York Dordrecht London

Printed on acid-free paper

Springer International Publishing AG Switzerland is part of Springer Science+Business Media (www.springer.com)

Preface

MEMS and Nanotechnology represents one of nine volumes of technical papers presented at the 2015 SEM Annual Conference and Exposition on Experimental and Applied Mechanics organized by the Society for Experimental Mechanics and held in Costa Mesa, CA, June 8–11, 2015. The complete Proceedings also include volumes on *Dynamic Behavior of Materials; Challenges In Mechanics of Time-Dependent Materials; Advancement of Optical Methods in Experimental Mechanics; Experimental and Applied Mechanics; Mechanics of Biological Systems and Materials; Mechanics of Composite & Multifunctional Materials; Fracture, Fatigue, Failure and Damage Evolution; and Residual Stress, Thermomechanics & Infrared Imaging, Hybrid Techniques and Inverse Problems.*

Each collection presents early findings from experimental and computational investigations on an important area within Experimental Mechanics, MEMS and Nanotechnology being one of these areas.

The MEMS and Nanotechnology fields are specialized scientific areas that involve miniaturizing conventional scale components and systems to take advantage of reduced size and weight and/or enhanced performance or novel functionality. These fields also encompass the application of principles ranging from the micron scale down to individual atoms. Sometimes these principles borrow from conventional scale laws but often involve new physical and/or chemical phenomena that require new behavioral laws and impart new properties to exploit. Studying how mechanical loads interact with components of these scales is important in developing new applications, as well as assessing their reliability and functionality. Establishing this symposium at the Annual Meeting of the Society for Experimental Mechanics provides a venue where state-of-the-art experimental methods can be leveraged in these endeavors.

The 2015 Symposium is the sixteenth in the series and addresses pertinent issues relating to design, analysis, fabrication, testing, optimization, and applications of MEMS and nanotechnology, especially as these issues relate to experimental mechanics of microscale and nanoscale structures.

It is with deep gratitude that we thank the Organizing Committee, Session Chairs, Authors and Keynote Speakers, Participants, and SEM Staff for making the 16th-*ISMAN* a valuable and unforgettable experience.

Auburn, USA
Wright-Patterson AFB, OH, USA

Barton C. Prorok
LaVern Starman

Contents

Chapter 1
Oxide Driven Strength Degradation of (1 1 1) Silicon Surfaces

Scott J. Grutzik and Alan T. Zehnder

Abstract Previous experiments have shown a link between oxidation and strength changes in single crystal silicon nanostructures but provided no clues as to the mechanisms leading to this relationship. Through a mix of atomic force microscope based fracture strength experiments, molecular dynamics modeling, and comparisons to past experiments we have shown that the previously described strength decrease is a result of oxidation induced roughening of an initially flat silicon (1 1 1) surface and we have shown this effect to be transient. In a fully developed native oxide the Si-SiO_2 interface flattens. As a result the strength recovers with some indication that it becomes slightly stronger than in its initially hydrogen terminated state.

Keywords Silicon • Silicon dioxide • Nanoscale strength • Nanoscale testing • Atomic force microscope

1.1 Introduction

Micro- and nanoelectromechanical systems (MEMS and NEMS) play an increasing role in many fields such as telecommunications, defense, microelectronics, and biomedical devices. These systems can be subjected to sustained loads, oscillating loads, dynamic shock loading, or a combination of all three. The small size of these devices make accurate predictions and measurements of strength and durability difficult. As a result they may be designed with overly large factors of safety with respect to mechanical failure. More efficient design will require accurate knowledge of the stochastic nature of fracture at these length scales.

For a variety of reasons, it is difficult to extrapolate macroscale material failure behavior down to the length scales required for MEMS and NEMS design. Components of these systems are often small enough that they contain few or no defects other than those introduced during processing. Also, because of their high surface to volume ratio, surface properties tend to play a larger role than volumetric properties. Nonetheless, it is possible to make some general predictions. From a continuum mechanics perspective, a smoother surface should give rise to a stronger structure. Any pit or ledge on the surface will act as a stress concentrator under loading and act as a potential nucleation point for fracture. It is also expected that if the surface can be maintained in a compressive state [1], strength should increase. The compressive stress would act to hold the faces of any existing cracks together, making it more difficult for cracks to propagate.

Alan et al. [2] have reported results that agree with the first of these expectations. Their results suggest that the surface roughness of nanoscale Si beams strongly affects their strength and that if the surface is smooth enough the strength can be close to the ideal strength derived from atomistic simulations. Alan et al. [3] have also reported results that are somewhat at odds with the second expectation that a compressive surface stress should increase fracture strength. Their data show a decrease in strength with increasing surface oxidation and, that if oxidation is prevented, the initially high strength is maintained. What makes this an unexpected result is that silicon dioxide grown on a Si surface grows in a compressive state. From our continuum mechanics reasoning oxidation should lead to an increase in fracture strength, not a decrease as the data suggests. The work presented here is part of an effort to understand the mechanisms behind how oxidation affects nanoscale strength.

S.J. Grutzik • A.T. Zehnder (✉)
Field of Theoretical and Applied Mechanics, Cornell University, Ithaca, NY 14853, USA
e-mail: atz2@cornell.edu

B.C. Prorok, L. Starman (eds.), *MEMS and Nanotechnology, Volume 5*,

1.2 Methods

1.2.1 Sample Preparation

Accurate measurement of mechanical properties such as stiffness or fracture strength on micro- and nanoscale samples is difficult. The small sizes and forces involved introduce challenges at every step of the process, from sample fabrication to gripping the specimen to measuring small forces and displacements. Several methods have been proposed that span a wide range of length scales and testing methods [4–7] Here we use a method developed by Alan et al. [2, 3] similar to that of Namazu et al. [4] in which a force is applied at the midpoint of a suspended single crystal beam via an atomic force microscope (AFM). A scanning electron micrograph of such a beam after testing is shown in Fig. 1.1. The test specimens are fabricated from Si (1 1 1) wafers using standard micro-machining techniques [8]. The long axis of the beam is parallel to the $\langle 1\ \bar{1}\ 0 \rangle$ direction.

The "dog bone" shape of the sample ensures that the fracture stress occurs reproducibly in the center of the beam, where the small width leads to highest stress. If the beam were uniform width, there would be stress concentrations at the attachment points which would complicate analysis. Figure 1.1 is post-fracture and the V shaped notch is the characteristic failure mode. Before testing, the center portion of the specimen would be uniform width. At failure, fracture occurs along (1 1 1) cleavage planes, creating the notch shape seen here.

1.2.2 Fracture Testing Procedure

After fabrication, an AFM is used to apply a force at the center of the beam, as shown in Fig. 1.2. This method has been previously used by Alan et al. The test has two parts. First, the beam is imaged by scanning the AFM cantilever over its surface to obtain its shape. Once the center of the beam is located, the cantilever is brought to this point and pushed straight down until the beam breaks. During deformation, the displacement applied to the base of the cantilever by the peizo in the AFM head, δ_{piezo}, and the output signal of the photodetector are recorded simultaneously until the beam fractures, as indicated by a sharp discontinuity in photodetector signal. Importantly, both the cantilever and the beam bend during this process. By repeating this deflection test on an effectively rigid area of the wafer surface, the photodetector signal can be

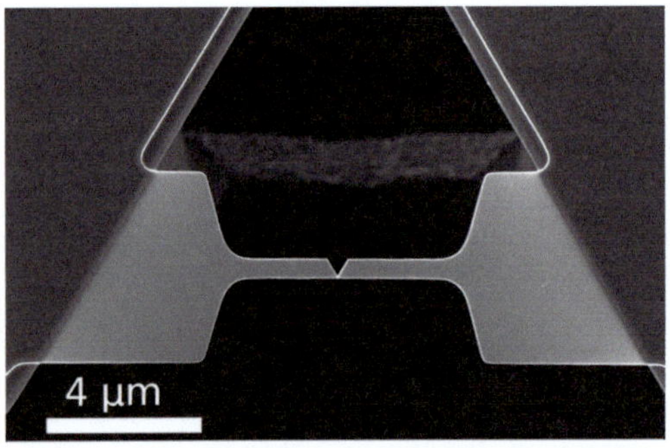

Fig. 1.1 Scanning electron micrograph of a fractured nanobeam. The V shaped notch is not part of the fabricated design but a characteristic of the failure mode

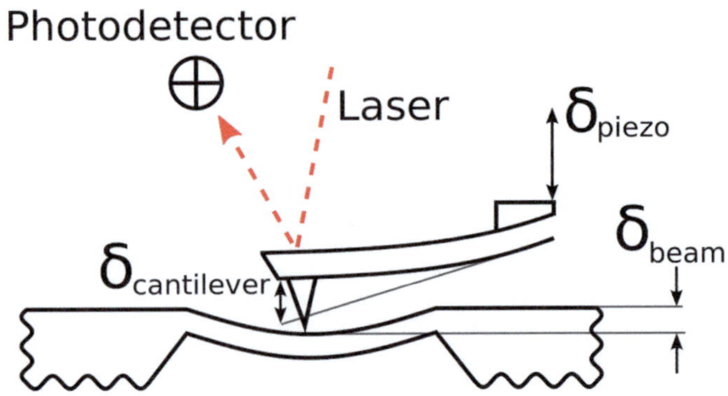

Fig. 1.2 Side view schematic of the AFM based fracture testing process

calibrated to the cantilever deflection, $\delta_{cantilever}$. The beam deflection, δ_{beam} is then calculated from $\delta_{beam} = \delta_{piezo} - \delta_{cantilever}$. If the stiffness of the AFM cantilever is known, $\delta_{cantilever}$ can be converted to the force F applied to the beam. This procedure quantifies the applied force F and beam deflection δ_{beam} at the time of failure. A variety of well characterized methods exist for measuring cantilever stiffness, most using mechanical or thermodynamic relations to determine stiffness [9, 10]. Due to the high stiffness required to generate enough force to fracture the Si beams used in the tests described here, the standard methods are not effective and a NIST traceable reference cantilever method was utilized instead [11].

Before the beam deflection and applied force at failure can be converted to maximum stress at failure, the thickness of the beam must be determined. To do this, we measure the resonant frequency of the beam by driving it inertially with a swept sine signal. Deflection during this measurement is transduced by optical interferometry [8]. The in-plane dimensions of the beam can be measured using SEM or the initial AFM scan which was used to locate the beam center. The thickness can then be calculated using the finite element method (FEM). The thickness in the FEM model is adjusted until the calculated resonant frequency agrees with the measured one. The validity of the FEM model can be verified by including higher order resonant frequencies in the thickness determination and by comparing the shape of the FEM generated load-deflection curve prior to failure to the experimental one. In both the physical and finite element beams, the long axis is aligned with the $[1\,\bar{1}\,0]$ direction. The material is assumed to be orthotropic with linear elastic modulii $C_{1111} = 166$ GPa, $C_{1212} = 64$ GPa, and $C_{4444} = 79.6$ GPa [12]. All strength testing was done on an Asylum MFP3D AFM (Asylum Research, Santa Barbara, CA). Finite element simulations were performed using ABAQUS (Dassault Systèmes, Waltham, MD) By applying the force at fracture to this FEM model, the fracture stress can be calculated.

1.3 Experiments and Results

1.3.1 Oxide Removal Experiment

The first step in determining the mechanisms by which oxidation is related to strength changes in Si nanostructures was detailed in our 2014 SEM proceedings paper [13]. In short, we prepared a set of nanobeam test specimens with H terminated surfaces. The H termination is unstable and oxidizes over a period of weeks to months. This natural oxidation is known as a native oxide and is well documented [14]. The beam specimens in question were allowed to oxidize over the course of 1 month while monitoring changes in strength. After this time, the oxide that had developed was removed with buffered oxide etch (BOE) and the strength was measured a final time. The results are shown in Fig. 1.3 with the red data point showing the strength after oxide removal. The Weibull strength is used as the characteristic strength measure and error bars are 95 % confidence intervals.

There were two possible outcomes of the oxide removal experiment. If the strength recovered after removing the oxide it would suggest that the presence of the oxide is behind the strength change. If the strength did not recover it would suggest the oxidation process caused an irreversible change in the silicon surface which resulted in a strength change. Figure 1.3 shows that strength does not recover after oxide removal. So what is the change that is occurring at the Si-SiO_2 interface? The most likely case is that oxidation is induces an atomic scale roughening of the beam surface. Neuwald et al. [15] used scanning

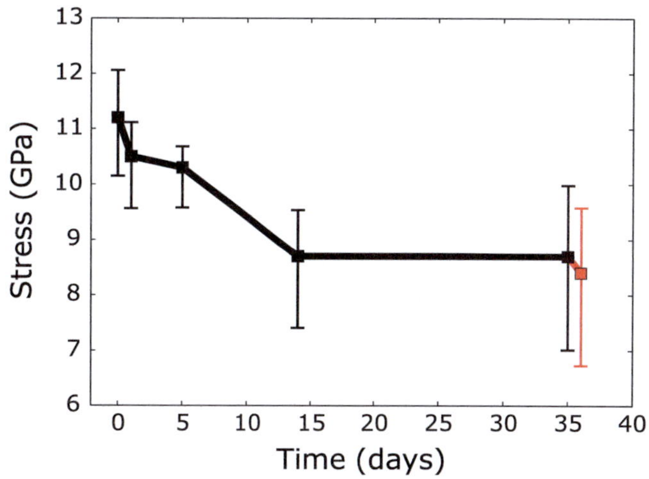

Fig. 1.3 Results of oxide removal experiment

tunneling microscopy to show that oxidation of an initially atomically flat Si (111) surface results in oxidation pits that form roughness on the order of single atomic steps. In addition, Alan et al. [2] showed that a similar level of roughness results in a similar level of strength change to that seen in Fig. 1.3.

1.3.2 Atomistic Modeling of Oxidation Induced Strength Changes

In an effort to support our hypothesis that atomistic scale roughness can lead to significant strength changes, we used molecular dynamics to calculate the effect of small surface steps and partial oxidation on strength. The open source molecular dynamics code LAMMPS [16] was used with the inter-atomic interaction modelled by ReaxFF van Duin et al. and Buehler et al. [17, 18]. A semi-infinite crystalline silicon region was modelled with an amorphous oxide region at the free surface. The amorphous oxide structure was created by initiating the oxide in an α-quartz structure, raising the temperature of the quartz region until it melted, then slowly cooling back to room temperature. The resulting SiO_2 structure is then annealed in an amorphous state [19]. The system was deformed in displacement control while allowing stresses in orthogonal directions to relax. This was done with surface steps of various heights as well as surfaces partially terminated with a thin oxide and the remainder remaining in H termination. The latter case simulates the effect of the nucleation phase of the oxidation process on strength. Figure 1.4 shows the geometry of the partially oxidized surface simulation.

Figure 1.5 is representative of the failure in all cases and shows a crack initiating at a step of 6 atomic planes and propagating normally to the tensile direction. Figure 1.6 shows failure strength with multiple step heights. Note that the relative strength change associated with small steps of 1–3 lattice planes is similar to the strength change seen in both our oxidation removal experiment and those by Alan et al. [2]. Figure 1.7 shows failure as a function of oxide nucleation fraction. Again, we see a similar level of relative strength change to that seen in experiments.

One unexpected feature of the failure modelled by ReaxFF seen in Fig. 1.5 is that the oxide appears to be deforming ductiley across the crack that is forming in the crystalline silicon. It is possible that this is an artifact of the MD simulation, either a fault of the ReaxFF formulation itself or maybe the annealing process left the oxide structure in a metastable,

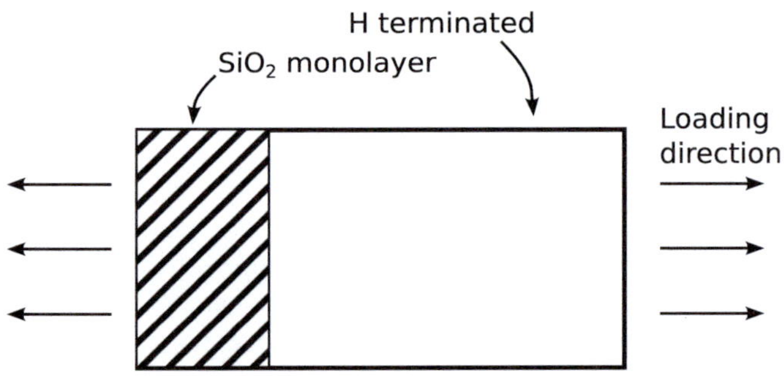

Fig. 1.4 Schematic showing the geometry of the molecular dynamics simulation calculating the effect of partial oxidation of a silicon surface on strength

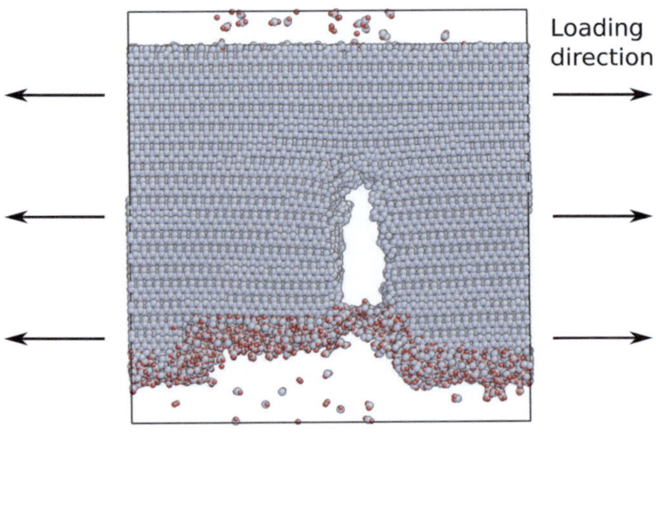

Fig. 1.5 A cross section of a molecular dynamics simulation of effect of small surface steps on strength of oxidized surface

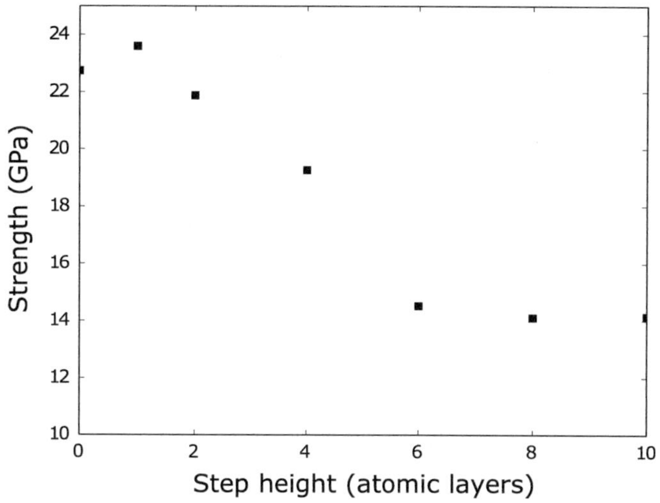

Fig. 1.6 Failure stress as a function of height of surface step as calculated by ReaxFF MD

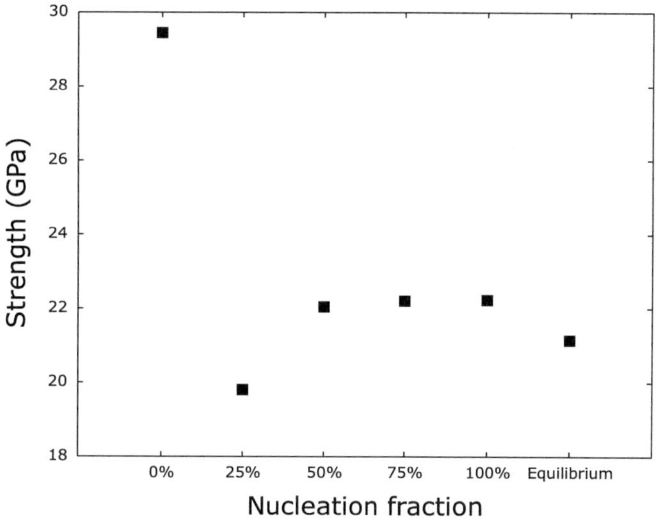

Fig. 1.7 Failure stress as a function of oxide nucleation fraction on a flat surface as calculated by ReaxFF MD

kinetically frustrated state that exhibits unphysical ductility. It is also possible that an oxide in this essentially two dimensional geometry really does deform in a ductile manner. If that is the case, it may have effects on strength beyond that of the oxidation induced roughening discussed earlier.

1.3.3 Artificial Oxidation Experiments

In order to determine whether the ductile deformation of a thin oxide as predicted by ReaxFF MD is physically plausible or merely a numerical artifact, a set of beam specimens was produced with oxides artificially grown to thickness in the range 2–12 nm. A native oxide will naturally grow to about 2 nm but this may take more than a year depending on atmospheric conditions. Two methods were used to artificially grow an oxide similar to a naturally occurring native oxide. A thin and a thick oxide were grown with a rapid thermal processor (RTP) [20] and a thin oxide was grown with ozone exposure at room temperature and atmospheric pressure [21]. In all cases, oxide thickness was measured with ellipsometry [22]. Failure strength was measured twice: in the initial H terminated state and after oxide growth. The results are shown in Table 1.1. In all cases, experimental error is about 10 %.

Table 1.1 Strength results for artificially grown thin oxides

	RTP thin (2.2 nm)	RTP thick (11 nm)	Ozone ($\sim$ 3 nm)
Pre-oxidation strength (GPa)	9.6	10.7	13.0
Post-oxidation strength (GPa)	10.4	11.5	12.5

All experimental uncertainties are about 10 %

In the cases of both the thin and thick RTP grown oxides, the strength after oxidation is higher than the initial strength. The strength increase, while consistent, is within experimental error in both cases. In the ozone assisted oxidation case strength drops slightly but well within experimental error. The data suggests that although strength drops with initial oxidation, with more well developed oxides strength is maintained and perhaps even increased. One potential cause for this small strength increase is the ductile behavior seen in the MD simulations. Compressive stresses in the oxide could also be suggested as a cause but experiments show the oxide film stress is much lower than the failure stresses seen here [1] and therefore not likely to be a major factor.

1.4 Summary

Atomic force microscope based experiments have shown that over a period of roughly 1 month the strength of initially H terminated Si nanostructures decreases by about 25 % with most of the change occurring within the first 2 weeks. This strength change can be shown to be a result of changes in the Si surface due to oxidation. By comparing the level of strength change to previous experiments which studied the effect of roughness on strength [2], and observing that scanning tunneling microscopy studies show that a similar roughness results from oxidation of an initially flat Si(1 1 1) surface [15], we conclude that the strength change is a result of atomic scaler roughness caused by oxidation. Molecular dynamics simulations corroborate this by showing similar strength decrease with small surface steps and partial oxidation of a simulated Si(1 1 1) surface. Molecular dynamics also suggests an unexpected ductile behavior in the oxide layer. Experiments with artificial oxides similar to fully developed native oxides show no strength decrease and perhaps a slight strength increase. We suggest that the initial oxide induced roughening is a transient effect of the oxidation front nucleating through each atomic layer and in equilibrium this oxidation front flattens. With the flattening, strength recovers with some indication that the fully developed oxide causes the structure to be slightly stronger than in its original state.

Acknowledgements This work was funded by the National Science Foundation (NSF) grant CMMI-0856488. SJG acknowledges support from the NSF IGERT program (DGE-0654193) It was performed in part at the Cornell NanoScale Facility, a member of the National Nanotechnology Infrastructure Network, which is supported by NSF Grant ECS-0335765 and made use of the Cornell Center for Materials Research Shared Facilities which are supported through the NSF MRSEC program (DMR-1120296). We would also like to express our gratitude to the Nanomechanical Properties Group at the National Institute of Standards and Technology for their aid in reference cantilever calibration.

References

1. Bhushan, B., Murarka, S.P., Gerlach, J.: J. Vac. Sci. Tehcnol. B **8**, 1068 (1990)
2. Alan, T., Hines, M.A., Zehnder, A.T.: Appl. Phys. Lett. **89**, 091901 (2006)
3. Alan, T., Zehnder, A.T., Sengupta, D., Hines, M.A.: Appl. Phys. Lett. **89**, 231905 (2006)
4. Namazu, T., Isono, Y., Tanaka, T.: In: The Thirteenth Annual International Conference on Micro Electro Mechanical Systems, pp. 205–210 (2000)
5. Tsuchiya, T., Tabata, O., Sakata, J., Taga, Y.: J. Microelectromech. Syst. **7**, 106–113 (1998)
6. Reedy, J.E.D., Boyce, B.L., Foulk, I.J.W., Field, J.R.V., Ohlhausen, J.A., de Boer, M.P., Hazra, S.S.: Predicting fracture in micron-scale polycrystalline silicon MEMS structures. Technical Report SAND2010-6701, Sandia National Laboratory (2010)
7. Gaither, M.S., DelRio, F.W., Gates, R.S., Fuller, J.E.R., Cook, R.F.: Scr. Mater. **63**(4), 422–425 (2010)
8. Wang, Y., Henry, J.A., Zehnder, A.T., Hines, M.A.: J. Phys. Chem. B **107**, 14270 (2003)
9. Ohler, B.: Practical advice on the determination of cantilever spring constants. Technical Report, Veeco Instruments Inc. (2007)
10. Veeco Instruments Inc.: Nanoscope V controller manual. (2006)
11. Grutzik, S.J., Gates, R.S., Gerbig, Y.B., Smith, D.T., Cook, R.F., Zehnder, A.T.: Rev. Sci. Instrum. **84**, 113706 (2013)
12. McSkimin, H.J., Andreatch, P.: J. Appl. Phys. **35**(7), 2161–2165 (1964)
13. Grutzik, S.J., Zehnder, A.T.: Oxide Driven Strength Degradation on (111) Silicon. In Fracture, Fatigue, Failure, and Damage Evolution, Volume 5 (pp. 113–117). Springer International Publishing. (2015)

14. Morita, M., Ohmi, T., Hasegawa, E., Kawakami, M., Ohwada, M.: J. Appl. Phys. **68**, 1272 (1990)
15. Neuwald, U., Hessel, H.E., Feltz, A., Memmert, U., Behm, R.J.: Appl. Phys. Lett. **60**(11), 1307–1309 (1992)
16. Plimpton, S.: J. Comput. Phys. **117**, 1–19 (1995)
17. van Duin, A.C.T., Dasgupta, S., Lorant, F., Goddard, I.W.A.: J. Phys. Chem. A **105**(41), 9396–9409 (2001)
18. Buehler, M.J., van Duin, A.C.T., Goddard, I.W.A.: Phys. Rev. Lett. **96**, 095505 (2006)
19. van Duin, A.C.T., Strachan, A., Stewman, S., Zhang, Q., Xu, X., Goddard, W.A.: J. Phys. Chem. A **107**, 3803–3811 (2003)
20. Lange, P., Bernt, H., Hartmannsgruber, E., Naumann, F.: J. Electrochem. Soc. **141**(1), 259–263 (1994)
21. Nakamura, K., Ichimura, S., Kurokawa, A., Koike, K., Inoue, G., Fukuda, T.: J. Vac. Sci. Technol. A **17**, 1275 (1999)
22. McCrackin, F.L., Passaglia, E., Stromberg, R.R., Steinberg, H.L.: J. Res. Natl. Bur. Stand. A **67A**(4), 363–377 (1963)

Chapter 2
In Situ TEM Nanomechanical Testing

Josh Kacher, Qian Yu, Claire Chisholm, Christoph Gammer, and Andrew M. Minor

Abstract Correlating the mechanical behavior of metals with the underlying defect mechanisms remains an outstanding challenge for the development of new material systems. In situ Transmission Electron Microscopy (TEM) nanomechanical testing provides an experimental technique whereby the behavior of defects such as dislocations and twins can be observed in real time while quantitatively correlating their behavior with an applied stress. This paper highlights recent experiments utilizing in situ TEM testing to investigate the behavior of twins and dislocations in FCC, BCC, and HCP materials. Examples of recently developed experimental approaches and future directions of in situ TEM nanomechanical testing are presented.

Keywords In situ TEM • Twinning • Nanowires • PLC effect • Nanomechanics

2.1 Introduction

The deformation of crystalline materials is dictated primarily by the interactions of defects such as dislocations and twins. How these interactions occur determines the mechanical properties such as strength and ductility. For example, precipitates and grain boundaries can inhibit the motion of dislocations and twins through the material, increasing the strength often at the expense of ductility. External influences such as temperature, irradiation, and corrosive environments affect these defect interactions and thus change the material properties. These influences on the mechanical behavior tend to be well understood empirically at the bulk scale, but the fundamentals of these interactions, the information needed to develop predictive, physically-based models, is lacking. Most approaches to mechanical testing of materials do not allow the visualization of defect interactions. Instead, samples are often thinned to electron transparency after completion of mechanical testing, facilitating post mortem characterization of the defect state, but leaving the defect processes to be pieced together from incomplete before and after images.

In situ nanomechanical testing via transmission electron microscopy (TEM) offers a pathway by which defect behavior and interactions can be observed directly during deformation. This approach facilitates the characterization of isolated dislocation/defect interactions. In addition, in situ TEM mechanical testing provides a practical platform to test nanostructured materials such as nanowires and nanoparticles. In these materials, the presence of free surfaces and interfaces introduces image forces that can alter deformation behavior and introduce defect interactions not observed in bulk materials. Increasing our understanding of such behavior will facilitate the more rapid development and application of novel material systems. Here, we describe examples of recently-developed in situ nanomechanical testing techniques and the insight gained into the mechanical behavior of small volumes resulting from their application.

J. Kacher • C. Chisholm • C. Gammer • A.M. Minor (✉)
National Center for Electron Microscopy, Molecular Foundry, Lawrence Berkeley National Laboratory, Berkeley, CA, USA

Department of Materials Science and Engineering, University of California, Berkeley, CA, USA
e-mail: aminor@berkeley.edu

Q. Yu
National Center for Electron Microscopy, Molecular Foundry, Lawrence Berkeley National Laboratory, Berkeley, CA, USA

Department of Materials Science and Engineering, University of California, Berkeley, CA, USA

Department of Materials Science and Engineering, University of Michigan, Ann Arbor, MI, USA

B.C. Prorok, L. Starman (eds.), *MEMS and Nanotechnology, Volume 5*,
Conference Proceedings of the Society for Experimental Mechanics Series, DOI 10.1007/978-3-319-22458-9_2

2.2 Main Text

2.2.1 Experimental Methods

This paper highlights examples for three different experimental approaches to in situ nanomechanical testing, illustrated in Fig. 2.1. The first approach, Fig. 2.1a, involves fixing a bulk-scale sample at either end and applying a global strain. An electron transparent region near the center of the sample provides a window in which to view defect interactions. Using this approach, defect interactions can be observed across tens to hundreds of microns but cannot provide quantitative stress and strain information. The second approach, shown in Fig. 2.1b, facilitates quantitative measurements of stress and strain. Here, an electron transparent strip of material is bonded to a push-to-pull (PTP) device using a platinum weld. The samples are generally prepared in a dual-beam focused ion beam (FIB), which also facilitates site-specific sample selection from the surface of a bulk sample or selection of a nanowire. The sample is loaded in tension by pushing the front of the PTP device. The third approach, shown in Fig. 2.1c, uses the same TEM holder as Fig. 2.1b, but with a freestanding sample such as a pillar or miniaturized dog bone sample for compressive and tensile loads, respectively. Again, quantitative stress and strain information is provided concurrently with direct viewing of the dislocation interactions.

2.2.2 Dislocation Processes Associated with the PLC Effect in Al 5754

Instability during straining of Al-Mg alloys, known as the Portevin-Le Chatelier (PLC) effect, limits the formability of these alloys and make it difficult to incorporate them into industrial applications [1]. These instabilities arise due to solute Mg atoms forming Cottrell atmospheres around glissile dislocations, temporarily arresting their progress until a sufficiently high stress level is reached. However, theoretical treatment of the PLC effect has struggled to reconcile the limited diffusion rate of Mg through the Al matrix with the observed mechanical behavior. Mechanisms such as pinning by forest dislocations and vacancy-enhanced diffusion of Mg have been proposed, but few experimental results capable of verifying these hypotheses exist in the literature [2]. Using the approach illustrated in Fig. 2.1a, Kacher et al. investigated the dislocation behavior in Al 5754 during in situ TEM deformation [3]. The motivation for this work was to characterize the effect of solute Mg on dislocation processes and relate this to the bulk mechanical behavior of Al-Mg alloys.

Figure 2.2 shows images taken from video displaying dislocations propagating from a crack tip during in situ TEM straining. Unexpectedly, the dislocations were seen interacting with obstacles dispersed throughout the matrix. The size of the obstacles was below the resolution of the in situ video, but their positions can be inferred from the dislocation behavior. An example of this is shown in Fig. 2.2a. Initially, a dislocation is seen pinned at two locations causing it to bow out. The dislocation first breaks free of one of the pinning points, causing it to partially wrap around the other. With further straining,

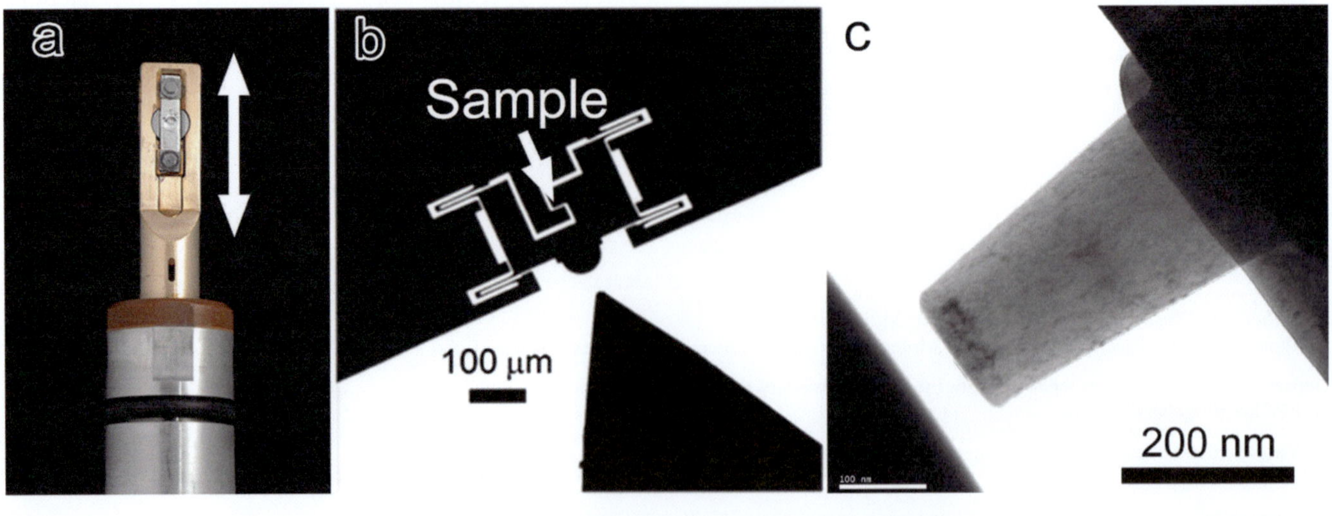

Fig. 2.1 Three approaches to in situ TEM nanomechanical testing. (**a**) Gatan model tension heating/straining holder. *Arrows* indicate tensile direction. (**b**) Push-to-pull device with a Hysitron PI-95 nanoindenter. (**c**) Nanopillar with flat punch (*bottom left*) near pillar surface

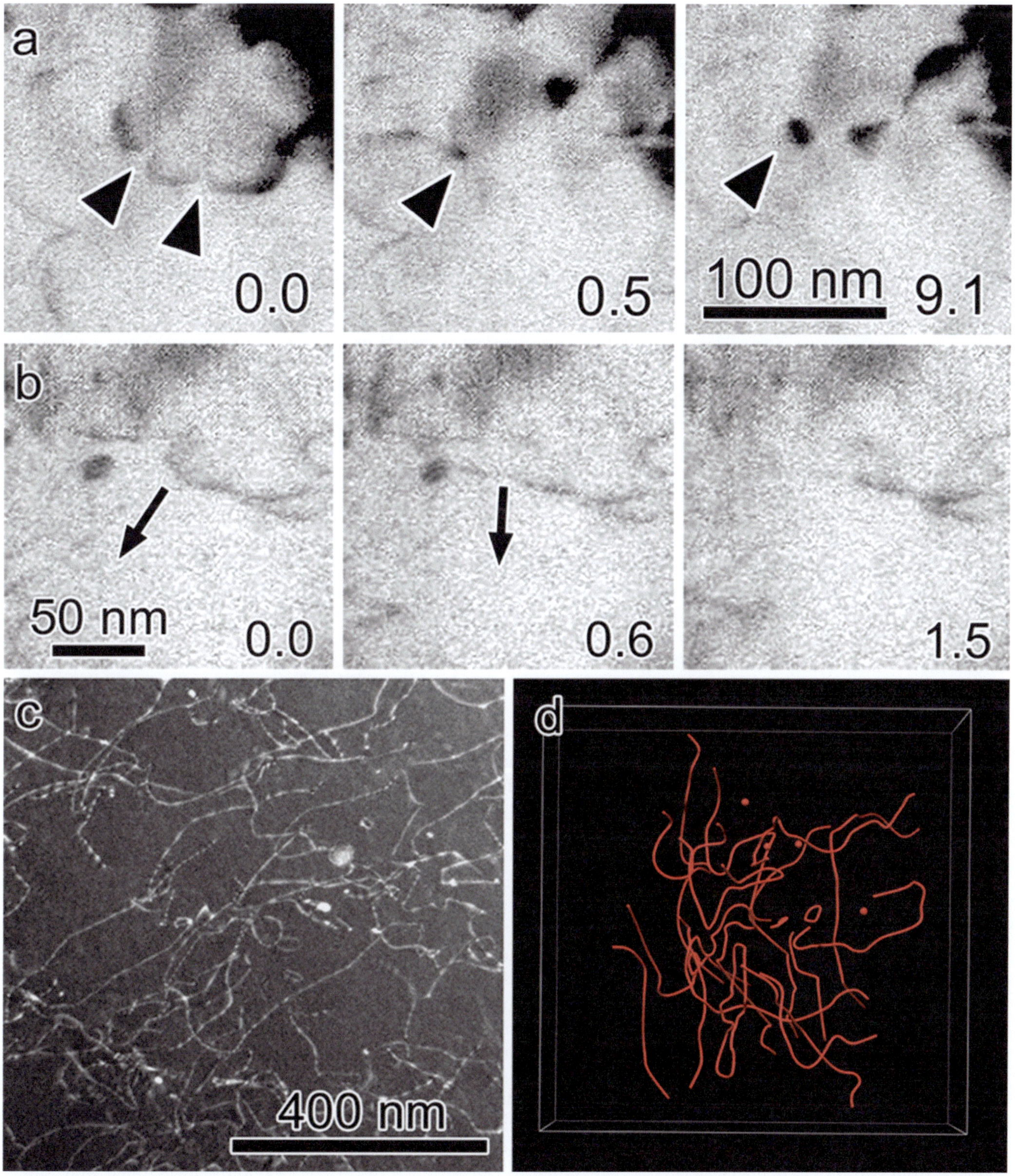

Fig. 2.2 (**a**) Dislocation loop formation recorded during in situ TEM straining. *Arrowheads* indicate obstacle locations. (**b**) Dislocation loop annihilation recorded during in situ TEM straining. *Arrow* indicates direction of dislocation motion. Experiment time given in each frame in seconds. (**c**) Weak-beam dark-field image of dislocations in deformed Al-Mg sample. (**d**) Three-dimensional model of dislocation structure seen in (**c**). The model construction was based off an electron tomogram reconstructed from the area shown in (**c**)

the dislocation completely encompasses the obstacle and pinches off, leaving behind a dislocation loop. Numerous dislocation loops were formed throughout the matrix via similar processes. These loops interacted with trailing glissile dislocations, commonly resulting in loop annihilation (see, for example, Fig. 2.2b).

The curvature of the dislocations can be related to the barrier strength of the obstacles by $\tau_s = \mu b/R$, where μ, the shear modulus, is given as 25.5 GPa for Al and b, the magnitude of the Burgers vector, is given as 0.29 nm [4]. R is the measured radius of curvature of the dislocation bowing around the obstacle immediately prior to the dislocation breaking free. As the dislocation character is unknown, this can only approximate the resolved shear stress. From the observed interactions recorded during in situ TEM deformation, ten discrete instances of pinned dislocations were analyzed giving an average barrier strength of 340 MPa. This value was compared to the average expected barrier strength of forest dislocations and was found to be over an order of magnitude higher, suggesting that their influence on the PLC effect is potentially larger than that which dispersed forest dislocations would provide [5].

To account for thin film effects present in in situ TEM experiments, dislocation structures formed during in situ deformation (Fig. 2.2c) were compared to samples prepared after ex situ tensile deformation of a bulk sample. Figure 2.2d is a three-dimensional dislocation model based off an electron tomogram of the area shown in Fig. 2.2c. The model shows that although there are many different dislocations present residing on various slip planes, no dislocation junctions have formed. One of the theories of the PLC effect depends on dislocation junctions forming as a pathway to Mg distribution to the glissile dislocations [2]. The dislocation model in Fig. 2.2d shows that the dislocation interactions occurring during deformation produce repulsive forces; critical information needed for modelling these interactions. The three-dimensional dislocation model also highlights the complex shapes of the dislocations. The dislocations kinks are seen to form in the matrix interior, similar to what was seen during the in situ deformation, suggesting that the observed interactions did not arise from thin-film effects.

The information combined from these different experiments provide a new mechanistic understanding of the PLC effect [3]. At the nano-scale Mg clusters form in the matrix, creating dispersed obstacles to further dislocation motion. These clusters cause the formation of Mg-enriched dislocation loops through pinch-off mechanisms and provide a pathway for Mg diffusion to dislocation cores as they are absorbed by glissile dislocations. The in situ TEM characterization was essential to understanding these dislocation processes, providing new insight into the PLC effect.

2.2.3 *Deformation in Nanowires*

The origin of size-dependent yield strength in small-scale samples has been linked to the effect of intrinsic and extrinsic length scales and the interactions of defects with interfaces or free surfaces [6]. Simulations developed by Weinberger and Cai have shown that surface effects on dislocations in body centered cubic (BCC) nanomaterials can result in a self-multiplication effect that leads to a strain softening effect at high stress that would not be present in larger volumes [7]. Experimental evidence of this phenomenon, however, has been difficult to obtain due to the challenges associated with testing nanoscale samples and directly verifying the. Chisholm et al. explored the deformation mechanisms of Mo nanowires via in situ TEM nanomechanical testing in order to characterize their failure mechanisms and identify factors leading to higher strength in BCC-structured nanomaterials [8].

A silicon PTP device loaded on Hysitron PI-95 picoindenter was used as a testing platform for in situ uniaxial straining experiments of the Mo nanowires (Fig. 2.3b). This device uses four laterally stiff springs and a micron-scale gap across which the sample can be placed with micromanipulators. Any mechanical vibrations of the indenter tip are dampened by the PTP, reducing mechanical noise. The load seen by the sample is calculated by subtracting the linear effect of the PTP springs from the raw load/displacement data. Engineering stress/strain is calculated using the initial cross-sectional area measured from TEM projections and the gage length as measured from scanning electron microscopy (SEM) images. In order to examine the local stress and strain in the location of a plastic deformation event, digital image correlation (DIC) can be used to measure the strain directly from the images using markers on the sample.

In order to reduce surface damage effects common to FIB-based sample preparation techniques, FIB-free samples were prepared from two Mo-alloy fibers (details of their synthesis can be found in [9]. The Mo-alloy fibers tested were single crystal with nominally square cross-sections around 200 nm and fiber axes along the <001> direction. The cross-sectional area was assumed to remain nominally square, based on post-mortem investigations of previous samples.

The results from one test that demonstrated a rapid strain softening event are shown in Fig. 2.3, showing video clips captured at a frame rate of 30 frames per second (fps) and the accompanied local true stress/strain curve. The wire initially had four dislocations that annihilated at the surface during loading, exhausting the sample of mobile defects. At a stress level of approximately 2.4 GPs, the wire was seen to fail suddenly and catastrophically. Interestingly, necking occurred at two

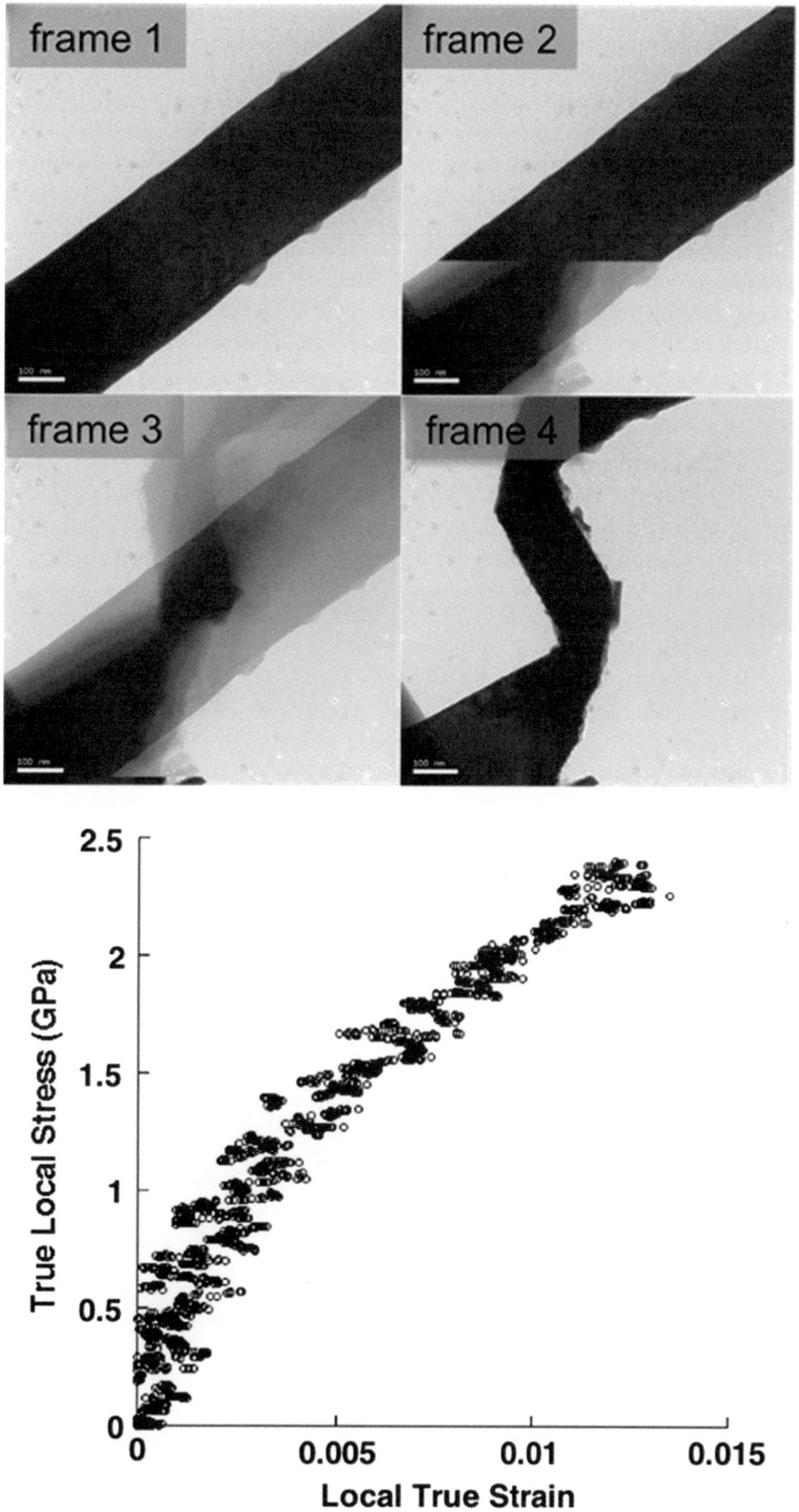

Fig. 2.3 Frames taken from in situ TEM straining video showing fracture of a Mo wire with associated local stress and strain information calculated using digital image correlation. The experiment was recorded at 30 frames per second, too slow to capture tof the first test, captured at a frame rate of 30 frames per second (fps) and accompanies with the true stress/strain curve, are shown in **a**. The wire initially had four dislocations that annihilated at the surface during loading, exhausting the sample. At a stress level of approximately 2.4 GPs, the wire was seen to fail suddenly and catastrophically. Interestingly, necking occurred at two distinct locations on the wire that, to the temporal resolution of the experiment, appeared to initiate simultaneously. This temhe actual strain softening event which occurs during the acquisition of frame 2

distinct locations on the wire that, to the temporal resolution of the experiment, appeared to initiate simultaneously. This temporal resolution was, however, insufficient to determine whether this failure was in reality simultaneous or sequential in nature. That is, the second necking point may be due to buckling of the nanowire associated with the initial failure event. In addition, no dislocation-based failure mechanisms were observed, preventing direct verification of the dislocation multiplication mechanisms predicted computationally. However, the true stress at time of failure can be used to compare with proposed failure mechanisms based off computer simulations. The self-multiplication model requires the stress in the sample to exceed the critical stress for loop formation, 1.0 GPa in this case. The observed failure of the Mo nanowire occurred at a true local stress of 2.4 GPa, offering supportive evidence for a self-multiplication-based failure mechanism.

These experiments highlighted both the usefulness of in situ nanomechanical testing, suggesting new failure mechanisms associated with simultaneous necking events and providing supportive information to previously proposed mechanisms. Current developments in dynamic TEM could be instrumental in overcoming some of the limitations in the temporal resolution, opening new possibilities in exploring deformation in nanomaterials.

2.3 Twinning in Nanopillars

Metals with HCP crystal structures such as Mg and Ti have a limited number of available slip systems and as a result they typically exhibit twinning as a prevalent mode of plastic deformation in addition to dislocation plasticity. Unlike the activation of a dislocation slip system, a single twinning event requires the interaction between different types of dislocations. For instance, according to the pole mechanism of deformation twinning in cubic crystals, the growth of a twin needs the cooperation of different dislocations in which the partial dislocations spiral around the pole (a screw dislocation), resulting in the shear of each atomic layers [10]. Twinning deformation is then sensitive to the availability of certain types of dislocations as well as the interactions between them. Although there is currently no systematic understanding of the twinning mechanisms in HCP structured materials because of the different crystal symmetry, the majority of current hypotheses for twin nucleation are reliant on dislocation-mediated mechanisms as well.

Using the in situ TEM nanomechanical testing apparatus shown in Fig. 2.1c, Yu et al. demonstrated that twinning deformation in HCP structured titanium and magnesium is directly related to the dislocation behavior. They tested singe crystal Ti-Al micron/submicron samples oriented for deformation twinning with millimeter size, micron size and submicron size pillars [11]. It was found that the formation of deformation twinning becomes more difficult in smaller samples in which the number of available "pole" screw dislocation is lower. Twinning deformation also follows the Hall–Petch rule; the critical stress for the formation of deformation twinning increases with decreasing sample size with the exponent in the Hall–Petch relation being approximately 0.92. Figure 2.4a shows the increase of flow stress with size. Interestingly, the mechanical properties of samples show significant change not only in strength but also in the plastic deformation behavior.

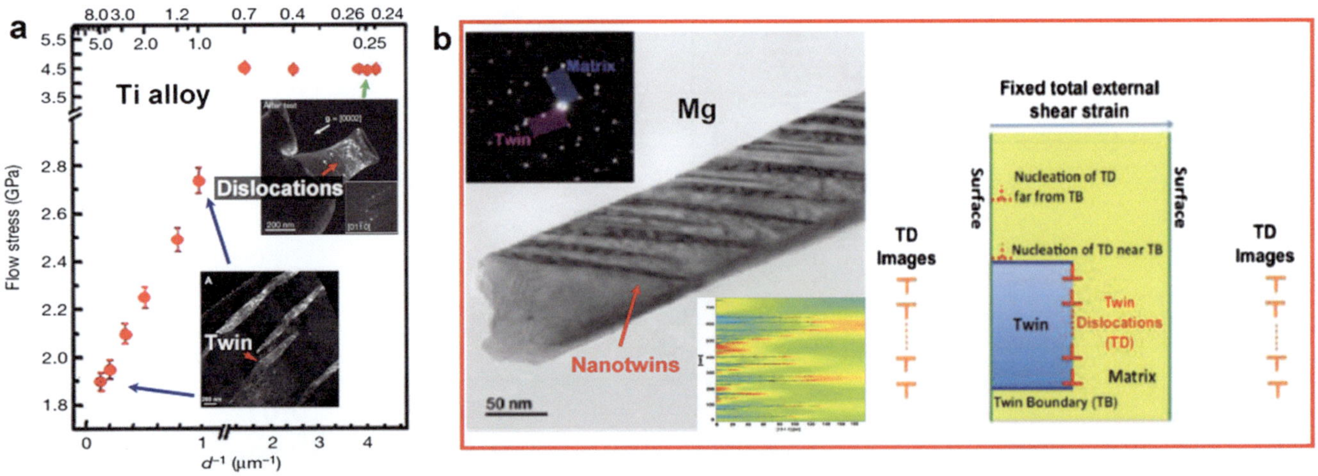

Fig. 2.4 (**a**) Plot of flow stress vs size in Ti-Al pillar tests. The TEM images *inset* show that the typical twinning structure observed in micron pillars and the dislocation structure observed in submicron pillars. The *left part* of (**b**) shows the typical nanotwin structure observed in tension and its corresponding diffraction pattern. The *inset* is the stress map generated by twinning dislocation interactions from Monte Carlo simulation. The *right part* of (**b**) is the schematic of the nanotwin mechanism proposed by Yu et al

By performing in situ TEM nanocompression tests on submicron samples, it was observed that twinning deformation was completely suppressed in the submicron-sized Ti-Al pillars even though the samples were oriented favorably for twinning deformation; the deformation mode changed from twinning-dominant to dislocation slip-dominant. The inset in Fig. 2.4a show the typical twinning structure observed in micron pillars and the dislocation structure observed in submicron pillars. The dislocation activity started from the top part of the specimen with the emission of multiple dislocation lines from the contact surface; the majority of the dislocations were <c + a> type. Since <a> dislocations are not mobile in this orientation and the mobility of <c + a> dislocations are relatively low, strong dislocation network formed immediately, significantly hindering further dislocation motion. With increasing stress, dislocation activity was seen throughout the sample. It was hypothesized that the lack of deformation twinning at submicron scale is due to the decreased population of screw dislocations that could act as pole dislocations as the sample volume decreased. As demonstrated above, with in situ TEM nanomechancial testing technique, one is able to gain direct experimental insights for the study of twinning mechanisms.

By combining in situ TEM nanomechancial testing and computational simulations to study plasticity in Mg, it was reported that the nucleation of deformation twins is directly related to the interaction between twinning dislocations [12]. In this study, in situ TEM tension, bending and compression tests were performed on single crystal Mg samples oriented for deformation twinning. In comparison to compression tests, tensile tests offer a more homogenous load distribution by minimizing the influence of sample taper and eliminating the sample surface/diamond tip interaction, thereby facilitating homogeneous nucleation of twins. Consequently, instead of the nucleation and growth of a dominant twin, the formation of nanotwins was observed in the in situ TEM tensile tests. The twins were a few nanometers thick and distributed throughout the gauge length of the tensile bars. The strength, work hardening and ductility were strongly enhanced as the high density of twin boundaries act as effective barriers for the motion of dislocations. Since Mg has a high stacking fault energy, this formation of nanotwins was unexpected. By using Monte Carlo simulations, it was discovered that the nucleation of multiple twins is promoted by the interaction of twinning dislocations in the previous twin. The interaction of the elastic fields of twinning dislocations at the tip of an existing twin would produce a high stress field located some distance away from this existing twin but a low stress field at the current twin boundary. As a result, the nucleation of a new twin from the high stress field is preferred rather than the thickening of the existing twin. This kinetic effect dominates the nucleation stage of deformation twinning, which is quenched down in the small samples in in situ TEM tensile test where the gauge length is small and the unloading is fast. Figure 2.4b shows the typical nanotwin structure observed in tension and its corresponding diffraction pattern. The inset is the stress map generated by twinning dislocation interactions from Monte Carlo simulation. The right part of Fig. 2.4b is the schematic of the nanotwin mechanism proposed by Yu et al.

2.4 Future Directions

The present paper demonstrates that in situ TEM not only helps us gain an understanding about deformation processes by imaging defect interactions in situ with nanoscale resolution, but also facilitates the simultaneous acquisition of stress and strain information in small volumes. This is made possible through recent developments of well-defined compression and tension geometries in combination with the use of new in situ TEM deformation holders that allow for the measurement of the load–displacement curve with high resolution. In addition to the global stress and strain, the local true plastic strain can be calculated using digital image correlation. Still, measurements of the elastic strain field around individual defects and other nanoscale features during deformation are not accessible using currently available experimental techniques. In the field of X-ray diffraction, there have been efforts to combine orientation mapping and strain mapping with in situ experiments, but only TEM has sufficient resolution to image individual defects [13].

Static measurements of the elastic strain field around defects have been demonstrated via TEM nanodiffraction [14]. This is possible by locally recording nanobeam diffraction patterns in a grid pattern over an area of interest in an electron transparent sample. Variations in the spacing between diffraction spots from pattern to pattern are directly proportional to changes in the d-spacing, which can then be related to the local elastic strain field. This approach combines high spatial resolution with a large field of view and has less stringent sample requirements than other strain mapping techniques. Once collected, the virtual apertures can be applied to the diffraction patterns to reconstruct dark-field images of the interrogated area using any of the collected diffraction spots [15]. Recently, the development of high-speed direct electron detectors paired with fast computing have made it possible to use diffraction mapping during in situ deformation without having to pause or stop the experiment. This experimental approach collects all available diffraction information during the in situ experiment, which can then be used to generate videos with optimized dark-field condition and compute the local transient strain with nanometer resolution. This is in contrast to current experimental approaches that require the selection of a single diffraction condition at the onset of an experiment and use that condition exclusively during testing. It is therefore envisioned that the concept of collecting all available signals during deformation using high-speed detectors and using data analysis to obtain quantitative information on local transient events, such as the movement of an individual defect, will open new research avenues in the field of in situ TEM deformation.

Acknowledgments JK supported by the National Science Foundation CMMI/MoM program under GOALI Grant 1235610. QY was supported by the General Motors Research and Development Center. CC was supported by the Center for Defect Physics, an Energy Frontier Research Center funded by the U.S. Department of Energy, Office of Science, Basic Energy Sciences. CG was supported by the Austrian Science Fund (FWF): [J3397]. All of the experimental work was performed at the Molecular Foundry, Lawrence Berkeley National Laboratory, which is supported by the U.S. Dept. of Energy under Contract # DE-AC02-05CH11231.

References

1. Yilmaz, A.: The Portevin-le Chatelier effect: a review of experimental findings. Sci Technol Adv Mater **12**(6), 063001 (2011) (16 pp.)
2. Picu, R.C., Zhang, D.: Atomistic study of pipe diffusion in Al-Mg alloys. Acta Materialia **52**(1), 161–71 (2004)
3. Kacher, J., Mishra, R.K., Minor, A.M.: Deformation-induced heterogeneous Mg distribution as a mechanism for the PLC effect in Al-Mg alloys. Philos. Mag. (2015)
4. Mompiou, F., et al.: In situ TEM observations of reverse dislocation motion upon unloading in tensile-deformed UFG aluminium. Acta Materialia **60**(8), 3402–14 (2012)
5. Madec, R., Devincre, B., Kubin, L.P.: From dislocation junctions to forest hardening. Phys. Rev. Lett. **89**(25), 255508–1 (2002)
6. El-Awady, J.A.: Unravelling the physics of size-dependent dislocation-mediated plasticity. Nat. Commun. **6**, 5926 (2015)
7. Weinberger, C.R., Cai, W.: Surface-controlled dislocation multiplication in metal micropillars. Proc. Natl. Acad. Sci. **105**(38), 14304–14307 (2008)
8. Chisholm, C., et al.: Dislocation starvation and exhaustion hardening in Mo alloy nanofibers. Acta Materialia **60**, 2258–2264 (2012)
9. Bei, H., George, E.P.: Microstructures and mechanical properties of a directionally solidified NiAl-Mo eutectic alloy. Acta Materialia **53**, 69–77 (2005)
10. Venables, J.A.: On dislocation pole models for twinning. Philos. Mag. **30**(5), 1165–1169 (1974)
11. Yu, Q., et al.: Strong crystal size effect on deformation twinning. Nature **463**, 335–338 (2010)
12. Yu, Q., et al.: The nanostructured origin of deformation twinning. Nano Lett. **12**, 887–892 (2012)
13. Jensen, D.J., et al.: X-ray microscopy in four dimensions. Mater. Today **9**, 18–25 (2006)
14. Hytch, M.J., Putaux, J.-L., Penisson, J.-M.: Measurement of the displacement field of dislocations to 0.03 A by electron microscopy. Nature **423**, 270–273 (2003)
15. Gammer, C., et al., Diffraction contrast imaging using virtual apertures. (2015) Submitted to Ultramicroscopy

Chapter 3
Poisson's Ratio as a Damage Index Sensed by Dual-Embedded Fiber Bragg Grating Sensor

C. Yılmaz, E. Akay, E.S. Kocaman, and M. Yildiz

Abstract Monitoring the health of glass or carbon fiber reinforced polymer under dynamic loading conditions is still a challenge. When metals are tested under dynamic loading conditions, usually a single crack is a source of the failure in the material. On the other hand, composite materials include several damage modes such as transverse cracking, delamination and splitting under the dynamic loading and all of them contribute ultimate failure of the material. Due to the complexity of the damage in composite materials, it is very hard to estimate health or damage state of composite materials. In this study, we propose the usage of a novel embedded biaxial Fiber Bragg Grating sensor system to track the evolution of Poisson's ratio which can be employed as a reliable damage index in composites. The fatigue experiments on specimen made of biaxial glass fiber infused with resin transfer molding system have shown that signal from novel embedded biaxial sensor system can be easily collected to evaluate Poisson's ratio. The current study also indicates that the evolution behavior of Poisson's ratio is consistent with other fatigue parameters such as temperature, force and strain energy that show very rapid change in the first region of fatigue with respect to number of cycles.

Keywords: Poisson's ratio • Glass fiber • Polymeric composites, biaxial FBG sensor, fatigue

3.1 Introduction

Composites, such as glass or carbon fiber reinforced polymer structures, have received significant interest due to their high specific strength and stiffness value for structural applications. Relatively high specific strength of composite structures makes them a prior structural material in applications where mass and operating costs are directly proportional. It is clear that when total mass of a vehicle such as airplane is reduced, less fuel consumption therefore more green vehicles can be achieved. On the other hand, when compared with metals, composite structures suffer from a lack of a well characterized fatigue life and damage mechanisms. Failure mechanism of metals during the fatigue is usually due to a single crack whereas composites can have multiple cracks which lead to a notable difficulty in the prediction of life of composites under fatigue.

Several different methods have been investigated to understand the accumulation of damage and damage state of composite structures under static and dynamic loading conditions. The degradation of elastic modulus was investigated in literature as a damage indicator for the composite materials [1]. The remaining useful life predictions with axial strain inputs were investigated to estimate remaining useful life (RUL) of glass fiber reinforced composite structures [2]. Apart from RUL approximation and usage of elastic modulus, several researches also examined the degradation of Poisson's ratio as a damage indicator [3–6]. Poisson's ratio can be used as a sensitive damage parameter to show damage state of composite materials under dynamic and static loading conditions. Measurement of Poisson's ratio for the quasi-static tests can be achieved by using strain gauges or extensometers. On the other hand, when dynamic loading conditions are considered, the evolution of Poisson's ratio becomes a nuisance due to the measurement methods such as strain gages and extensometers.

Poisson's ratio is influenced by the applied longitudinal strain or transverse crack density [7, 9]. Paepegem et al. [7] investigated evolution of Poisson's ratio of composite materials with respect to applied longitudinal strain and observed interesting behavior in the Poisson ratio as a function of applied strain such that the Poisson's ratio indicates a descending

C. Yılmaz • E.S. Kocaman • M. Yildiz (✉)
Faculty of Engineering and Natural Sciences, Advanced Composites and Polymer Processing Laboratory (AC2PL), 34956 Istanbul, Turkey
e-mail: meyildiz@sabanciuniv.edu

E. Akay
Faculty of Aeronautics and Astronautics, Astronautical Engineering, Istanbul Technical University, 34469 Maslak, İstanbul, Turkey

B.C. Prorok, L. Starman (eds.), *MEMS and Nanotechnology, Volume 5*,
Conference Proceedings of the Society for Experimental Mechanics Series, DOI 10.1007/978-3-319-22458-9_3

trend as a function of the applied strain. Smith et al. [9] modeled evolution of Poisson's ratio with shear-lag theory under static loading conditions and correlated their analytic model with experimental results. The evolution of Poisson's ratio is plotted as a function of transverse crack density and it was shown that with increasing transverse crack density, Poisson's ratio indicates a decreasing trend. F. Gao et al. [10] also studied the effect of delamination on the variation of Poisson ratio.

Not being limited to composite materials, the evolution of Poisson's ratio is a significant material property for biomedical research. Pidaparti et al. [8] investigated the change in Poisson's ratio during stress controlled dynamic loading for cortigal bone research. He observed that reduction of Poisson's ratio is more than the reduction of longitudinal modulus of elasticity with respect to normalized number of fatigue cycles. The difference in the reduction between the Poisson's ratio and elastic modulus is attributed to accumulation of transverse micro-cracking on the cortigal bone. It was reported that the accumulation of transverse damage is three times higher than the accumulation of longitudinal damage for a studied cortical bone under fatigue loading. Hence, Poisson's ratio was considered to be an effective damage index for biologic parts since it embodies both axial strain and transverse strain as seen in Eq. (3.1);

$$\nu_{12} = -\varepsilon_{\mathrm{y}}/\varepsilon_{\mathrm{x}} \tag{3.1}$$

When considering evolution of Poisson' s ratio for both dynamic and static loading conditions, measurements have always been done by using strain gauges or extensometer which is common point of above described researches. Both types of sensors collect the strain from surface of the test specimen and cannot be part of the structure. In addition, strain gage and extensometer are sensitive to electromagnetic interference, and thus, cannot used in environments with high electromagnetic field. To circumvent the drawbacks of strain measurements with surface mounted sensors systems, one promising approach would be the utilization of embedded sensor system. One possible candidate for embedded sensor system is Fiber Bragg Grating (FBG). FBG is a part of fiber optic cable that consists of periodically varying refractive index on the core of a fiber along the fiber direction. It can be thought as a crystal lattice that behaves like stop-band filter [11]. A small portion of the electromagnetic spectrum is reflected back while others pass the grating section. Reflected portion of incident electromagnetic spectrum is centered at a wavelength known as Bragg wavelength, λ_B. The Bragg wavelength λ_B is given by

$$\lambda_{\mathrm{B}} = 2n_{\mathrm{eff}}\Lambda \tag{3.2}$$

where Λ is grating period and n_{eff} is the effective refractive index of FBG. When strain or temperature of fiber varies, it changes the grating period and effective refractive index in grating part which causes a shift in the Bragg wavelength λ_B. Change in the Bragg wavelength can be determined using the following equation

$$\frac{\Delta\lambda_B}{\lambda_B} = (\alpha + \xi)\Delta T + (1 - \rho_e)\varepsilon \tag{3.3}$$

where α and ξ are the thermal expansion coefficient and thermo-optic coefficient of fiber core respectively. ΔT is the change in the temperature in grating region and ρ_e denotes effective photo-elastic cosnstant of fiber core, ε is the axial strain of grating region.

In literature, the evolution of Poisson's ratio has been investigated with respect to applied longitudinal strain [4–6] and proposed that the Poisson's ratio is closely related with the transverse crack density in the composites [3]. Nevertheless, when a real life structures and applications are considered, it is important and critical to track change in Poisson's ratio with respect to fatigue life or number of cycles of composite whereby the fatigue life and the damage state of composite can be monitored more effectively and conveniently. Therefore, in this study, we have monitored the evolution of Poisson's ratio of composite materials subjected to fatigue loading as a function of fatigue cycle number using embedded axial and transverse FBG sensors. To our best knowledge, the evolution of Poisson's ratio of glass fiber reinforced epoxy composite using embedded biaxial FBG sensors has not been presented in the open literature yet. Moreover, axial force on the composite, expended strain energy, and temperature change of composite during fatigue are also investigated along with evolution of Poisson' s ratio to show that Poisson's ratio measured by embedded FBG sensors is a reliable quantity which can be used as a damage index.

3.2 Materials and Method

3.2.1 Composite Materials

As reinforcement, 329 gsm biaxial (0/90) E-glass stitched fabric (Metyx, Turkey) with 177 gsm along the warp (0°) direction (which is the loading direction of the specimen) and 142 gsm along the weft (90°) direction (which is perpendicular to the loading direction) and 10 gsm stich is used. As a matrix material, Araldite LY 564 epoxy and XB3403 hardener system purchased from Huntsman is utilized. Glass fiber reinforced epoxy composite plates were manufactured by using Resin Transfer Molding system with a special design that enabled the embedding of fiber optic sensors into the composite plate during fabrication. Resin was injected to the mold by applying external pressure up to 2 bar. Composites samples underwent an initial cure process at 65 °C for 24 h then a post cure for another 24 h at 80 °C. The configuration of the composite plates is $[90/0]_{6S}$. Composite samples with fiber optic sensors were cut with a water cooled diamond saw. Figure 3.1 presents the drawing and picture of a specimen. The section of the panel including the FBG sensor is cut into L-shape so that the specimen can be clamped by the grips of an MTS fatigue machine without damaging the egress region of the FBG sensor. In Fig. 3.1c is shown a composite specimen mounted on the hydraulic grips of the fatigue test system

3.2.2 Equipment and Test

All fatigue tests were performed using a servo-hydraulic MTS 322 test frame with a MTS FlexTest GT digital controller and a load cell of ±50 kN. Fatigue tests on composites specimens with embedded biaxial FBG sensors were conducted under displacement control using a linear variable differential transformer (LVDT) as a control sensor integrated into the load cell.

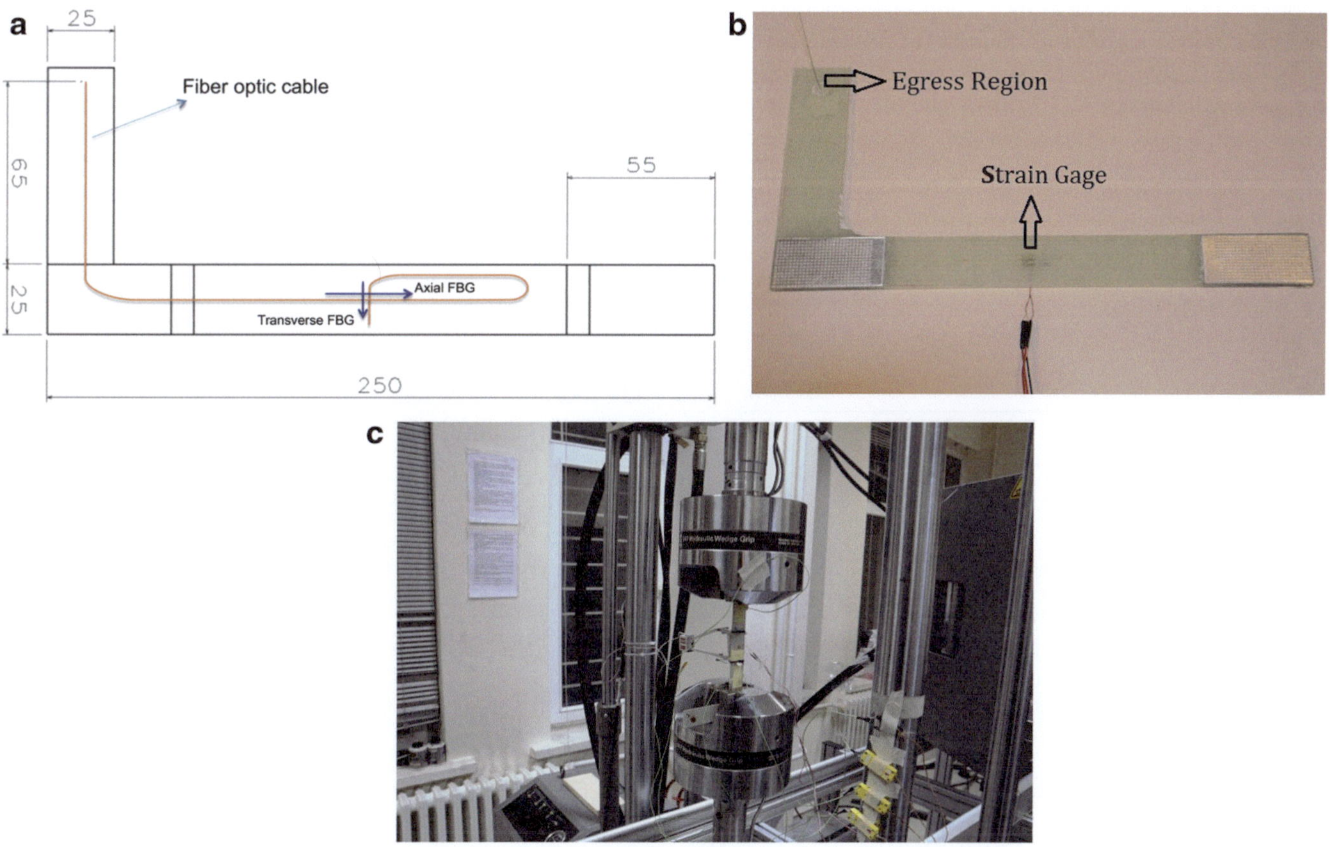

Fig. 3.1 (**a**) Drawing of fatigue test specimen with an embedded biaxial FBG sensor, (**b**) fatigue test specimen with embedded FBG sensor and strain gage, and (**c**) fatigue test on the specimen

Fatigue tests were performed under tension-tension mode with a constant strain amplitude of $\varepsilon_{max}/\varepsilon_{max} = 0.6$ at a frequency of 4 Hz where the ultimate strain of the composite is 16,200 με. Multiplexed biaxial FBG sensors are embedded in specimens as shown in Fig. 3.1 where the both axial and transversal FBG sensors have a gage length of 1 mm and have a center wavelength of 1550 nm and 1540 nm, respectively. A Micron Optics SM230 model interrogator was used to acquire the FBG signals during the fatigue experiment with Micron Optics Enlight Software. Fiber optics sensors are procured from Technesa. Strain gage and thermocouple data were collected concurrently by National Instruments NI SCXI-1000 main chassis with NI SCXI-1520 and NI SCXI-1314 card, respectively. LabVIEW software was used for data acquisition from K-type thermocouple and strain gauges.

3.3 Results and Discussions

Before a fatigue test, a test specimen was loaded in a quasi-static cyclic manner four times to calibrate FBG sensor as well as eliminate any slipping of the sample from the grips during fatigue testing until reaching the strain ratio of 0.6. FBG sensors were calibrated by different sensor systems using the data obtained in the fourth loading cycle and results are presented in Table 3.1. As seen in Table 3.1, the calibration coefficient of axial FBG sensor can be different for extensometer and axial strain gage. This difference is most likely due to the difference in the gage length between the strain gage and extensometer. Upon paying a close attention on Table 3.1, one may note that the calibration coefficient of transverse FBG is higher than axial FBG.

In the displacement controlled fatigue tests, the entire gage length of the specimen is considered by the LVDT sensor thereby taking into account strain relaxation due to transverse cracks or other damage modes along the full gauge length of the test specimen. In Fig. 3.2a, b are given the axial and transverse strains acquired by an embedded biaxial sensor system respectively for a certain range of certain cycles where one can immediately notice that both cyclic signals are free of any noise as is typical for a strain gage based acquisition. It is worthy of reiterating that with the careful placement of a multiplexed sensor system into composite structures, biaxial strains can be reliably measured, which is not possible with electric based strain sensors, namely, extensometer and strain gage system.

Fatigue in fiber reinforced polymer matrix composites are characterized by three distinct stages. Namely, in the first stage comprising the first 15–25 % of fatigue life, the rapid formation and interconnection of matrix cracking causes a sharp, non-linear decrease in stiffness. The second stage accounts between 15 and 20 % to 90 % of the fatigue life where there is a

Table 3.1 Calibration coefficients of axial and transverse FBG sensors

Axial Extensometer/Axial FBG (με/nm)	Axial Strain Gage/Axial FBG (με/nm)	Transverse Strain Gage/Transverse FBG (με/nm)
773	677	1209

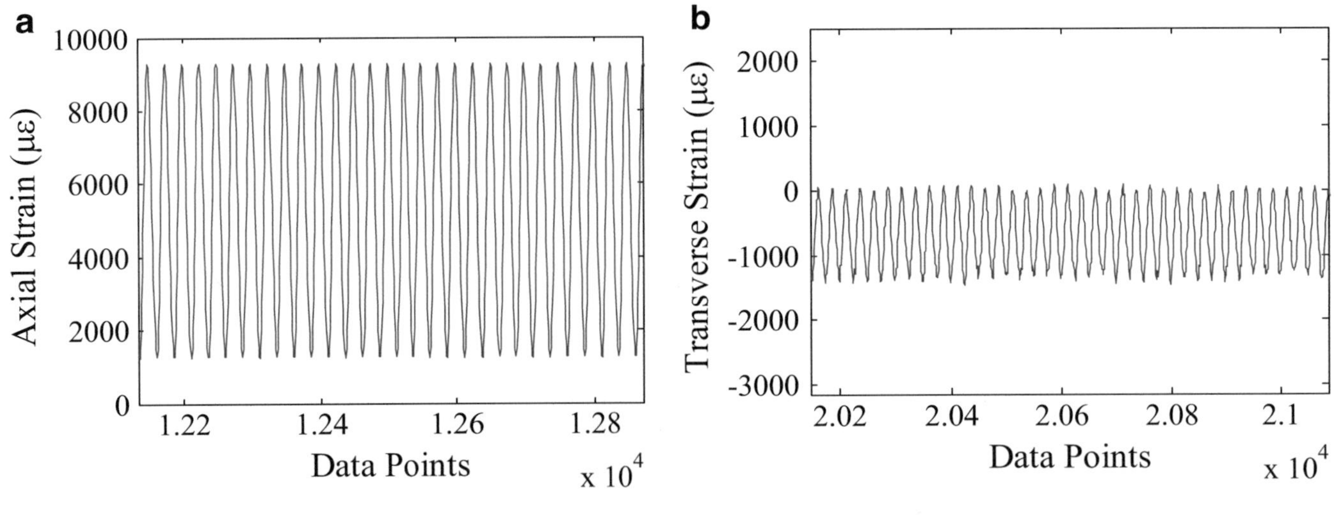

Fig. 3.2 Values of strains gathered by, (**a**) axial FBG, (**b**) transverse FBG

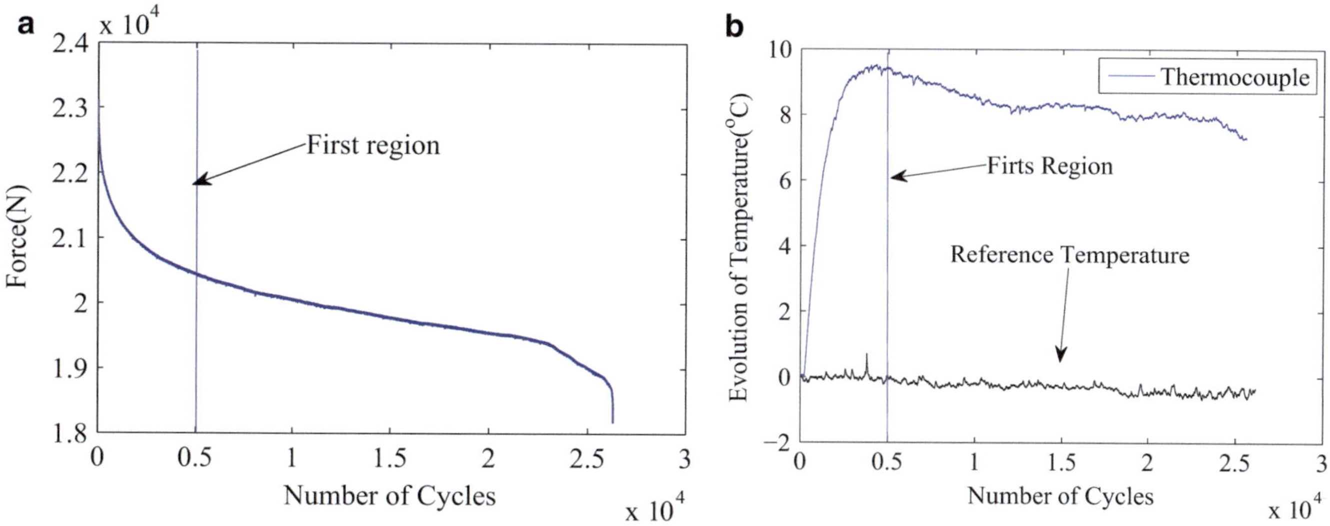

Fig. 3.3 The evolution of force on the specimen (**a**) and the variation of the temperature measured on the surface of the specimen (**b**) in the course of fatigue experiment

gradual, and linear decrease in stiffness, which is attributed to crack propagations, fiber debonding and delamination. The final stage is differentiated by a sharp nonlinear decrease in stiffness due to the plurality of fiber breakages. Figure 3.3a shows the evolution of applied fatigue force during the test which possesses three distinct stages as expected since stiffness degradation or decrease in the applied force are equivalently related one another for the fatigue experiment conducted at constant strain amplitude as is the case for the current study. In this experiment, the first fatigue phase spans over about 5000 cycles (corresponding to nearly to 20 % of the failure cycle) in which the force on the test specimen continuously decreases as expected since the fatigue experiment is performed under constant displacement and as a result of fatigue damage, in each cycle, less and less force is required to reach the same initial displacement. Recall that the cause of such a noticeably sharp decline has been associated with the matrix cracking, where in this particular loading configuration, the damage are likely to be dominated by transverse cracks. After the 5000 cycle, the decent in the force continues albeit with a smaller rate and almost follows a linear trend up to the last 1000 cycles with respect to the first region where this stage is referred to as the second phase of the fatigue. Then rather close to the failure which generally corresponds to the 10 % of the overall fatigue life, the third stage is entered which is characterized by drastic drop in the applied force.

It is known that when materials such as metals, and polymeric composites are exposed to repetitive loading, they heat up due to the fact that the portion of the mechanical strain energy on the fatigue specimen is dissipated as heat which might be contributed by inherent frictions within any material system (even in the absence of any defect) as in the case of viscous dissipation in fluidic systems, and also by frictions between the surfaces of fatigue induced cracks (hereafter called as induced friction). This heating phenomenon is referred to as autogenous heating. In Fig. 3.3b is seen that the temperature of the specimen under fatigue loading increases by around 10 °C in a rather short cycle number, which is roughly 20 % of the total cycle number for failure. This length of this thermal first phase is in agreement with that of the first fatigue phase. The temperature of the specimen gradually decreases and then levels off. This region is called as the second thermal phase, and corresponds to the second fatigue phase in Fig. 3.3a. The reason behind the temperature drop is that the rate of heat generation is well balanced by the rate of heat removal to the surrounding through conduction, convection and radiation although at such small temperature range radiative heat transfer is prudently negligible. Furthermore, noting that convective heat flux in the direction perpendicular to the surface of the specimen ($q_c = h\Delta T$ with $\Delta T = (T - T_\infty)$) augments due to the increase in ΔT where T is the temperature of the surface and T_∞ is the ambient temperature. Moreover, as the force decreases, less mechanical work would be done on the system so that the amount of heat dissipation will naturally be smaller.

Alternatively, the data leading to Fig. 3.3a can be presented as expended strain energy versus the fatigue cycle number of test specimen as seen in Fig. 3.4 where the expended strain energy in each cycle is calculated as $U_n = P_n\delta_n/2$. Here, U_n is expended strain energy for nth cycle, P_n is the maximum applied force for nth cycle and δ_n is the maximum displacement for the nth cycle. In Fig. 3.4, the rate of expended strain energy follows very rapid increasing trend up to the boundary of the first fatigue region. In the first region, the expended strain energy behaves like a logarithmic function. After the first region, increase in the expended strain energy becomes linear with respect to the cycle number. The logarithmic and linear increase trend in the first and second region respectively can be attributed to the rate of transverse crack formation given that in the

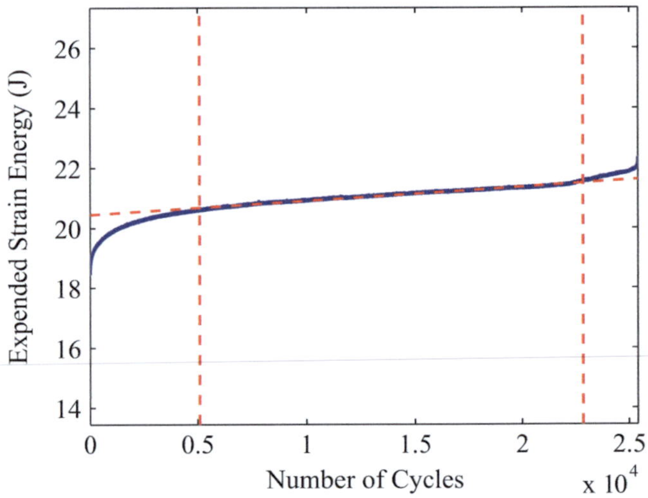

Fig. 3.4 Expended strain energy versus number of cycles

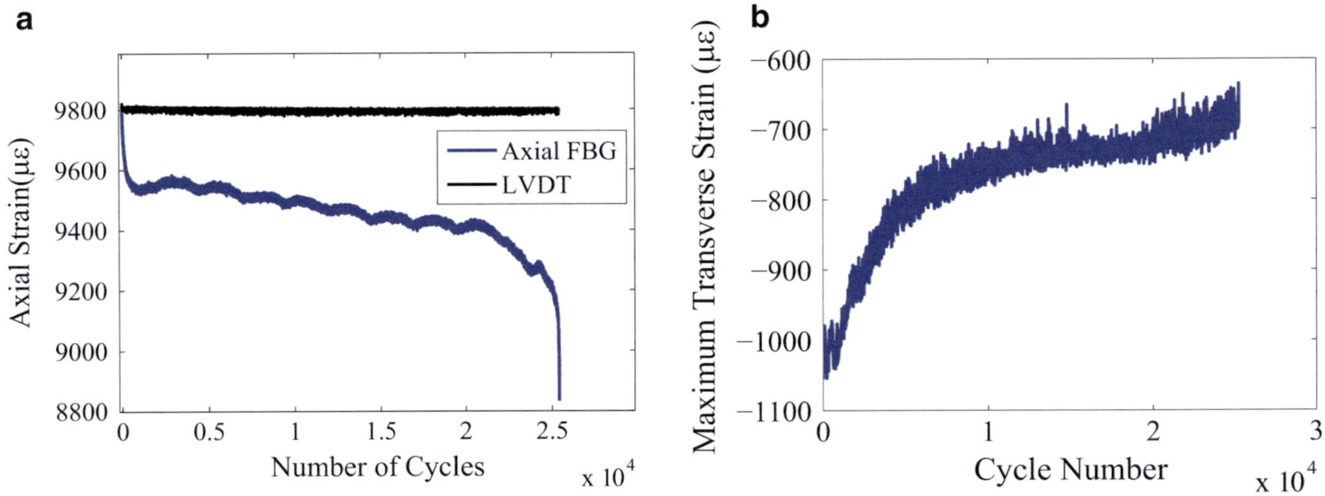

Fig. 3.5 (**a**) Peak values of cyclic axial strain obtained converting the diplacement recorded by LVDT into the strain through using calibartion coefcient and measured by an axial FBG sensor, and (**b**) the peak values of transverse strain measured by transversely embedded FBG sensor

first region, the rate of transverse crack formation is very high while for the second region, the rate of the transverse crack formation comes at a steady state resulting with a very low crack formation. After the first region, other damages modes start to contribute deformation of material. These damage modes are delamination and splitting that cause a linear increase in expended strain energy.

As seen in the Table 3.1, the method of strain measurement on the structure or material is very sensitive to the type of sensor used, namely, which can be extensometer, LVDT, strain gauge, or FBG sensor. As stated, an FBG sensor is an embedded sensor, and sense the strain locally due to its rather small gage length (i.e., varying between 1 mm and 10 mm or longer) oriented along the loading axis of test specimen 1 mm. An FBG sensor embedded into the test specimen during the manufacturing step of the composite can act as a global strain measurement device provide that transverse cracks do not disturb, recalling that the axial FBG sensor written on the fiber optic cable lies along the entire gauge length of the fatigue specimen. In the absence of cracks in the structure, the fiber optic cable experiences the same strain with the entire gauge length of the fatigue sample. Noting that the fiber optic cable and FBG sensor are on the same line, all the axial strain must be transferred from the fiber optic cable to the FBG sensor if there is no disturbance such as the transverse crack or delamination. However, the formation of transverse crack is ineluctable for cross ply-composites laminates under fatigue testing. When the transverse crack coincides with the gauge section of the axial FBG sensor, a local strain relaxation occurs in the region of sensor and this strain relaxation can be sensed by axial FBG sensor as shown in Fig. 3.5, hence resulting in the reduction on the measured axial strain value. Although the strain measured by LVDT was constant during the fatigue

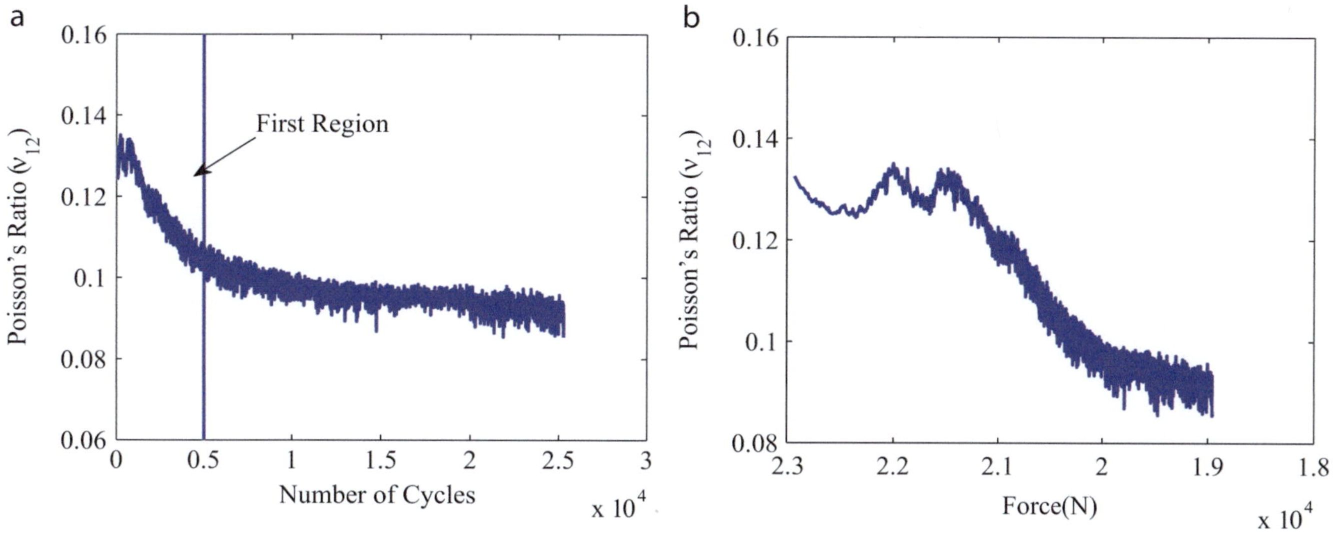

Fig. 3.6 The evolution of poisson's ratio; (**a**) with respect to cycle number, and (**b**) with respect to maximum force

testing, the axial FBG sensor follows a decreasing trend with small fluctuations that can be seen in the blue line in Fig. 3.5a. These small fluctuations might have been due to the local strain relaxations in the region of FBG gauge length. In the last 1000 cycle, the strain measured by the axial FBG experiences a rapid decrease which indicates the end of the fatigue life of test specimen. On the other hand, the transverse strain measured by transversely positioned FBG sensor as shown in Fig. 3.5b does not have a rapid change within the last 1000 cycle. As mentioned earlier, FBG is an embedded sensor and hence can sense damage state of the structures more precisely than the conventional sensors such as strain gauges, extensometer and LVDT.

It is obvious that the accumulation of damages with respect to cycle number has a significant effect on the measured fatigue parameters such as force, expended strain energy and surface temperature of samples. Above described fatigue parameters are measured by externally attached sensor system on the material. On the other hand, the Poisson's ratio which is an important engineering constant is measured by embedded biaxial FBG sensor system without deteriorating the integrity of composite material. Cycle by cycle analysis of Poisson's ratio calculated by embedded biaxial FBG sensor system can be seen in Fig. 3.6a, b. To calculate the Poisson's ratio, both axial and transverse FBG sensors was calibrated by biaxial strain gages just before the fatigue test by applying the previously described static tensile load up to 0.6 strain ratio. Being similar to the other fatigue parameters, Poisson's ratio indicates a rapid change in the first region of fatigue test. When the first phase of fatigue test ends, Poisson's ratio no longer changes significantly with respect to cycle number as seen in Fig. 3.6a. The rapid change in Poisson's ratio with respect to cycle number is consistent with other measured fatigue parameters such as temperature, force, or expended strain energy. The reduction in the Poisson's ratio can be linked to the decreasing trend of peak values of transverse strain in the so called first region as shown in Fig. 3.6b. The Poisson's ratio is also plotted with respect to applied force on the material and is observed to decrease with the increase in the applied load.

3.4 Conclusion

This work has aimed to monitor evolution of Poisson's ratio of glass fiber reinforced polymers (GFRP) subjected to tension-tension fatigue loading using embedded Fiber Bragg Grating sensors. It was shown that Poisson's ratio indicates a decreasing trend during fatigue loading. As result of fatigue loading, the GFRP specimens degrade continuously due to the creation of damage therein. The current study shows that the Poisson's ratio can be used an effective damage index measured by embedded optical sensor system. The change in Poisson ratio measured by dual embedded optic sensor system is also consistent with other measured fatigue parameters such as force, temperature, and expended strain energy.

Acknowledgements The authors gratefully acknowledge the funding provided by The Scientific and Technological Research Council of Turkey (TUBITAK), and Ministry of Science, Industry and Technology of Turkey for the project, 112M357, and 01307.STZ.2012-1, respectively.

References

1. Highsmith, A.L., Reifsnider, K.L. Stiffness-reduction mechanisms in composite laminates, damage in composite materials, ASTM STP775. In: Reifsnider, K.L. (ed.), American Society for Testing and Materials, pp. 103–117
2. Keulen, C. J., Akay, E., Melemez, F. F., Kocaman, E. S., Deniz, A., Yilmaz, C., Yildiz, M., Suleman, A. (2014). Prediction of fatigue response of composite structures by monitoring the strain energy release rate with embedded fiber Bragg gratings. Journal of Intelligent Material Systems and Structures, 1045389X14560358.
3. Gao, F., Boniface, L., Ogin, S.L., Smith, P.A., Greaves, R.P.: Damage accumulation in woven-fabric CFRP laminates under tensile loading: Part 1. Observations of damage accumulation. Composit. Sci. Technol. **59**(1), 123–136 (1999)
4. Surgeon, M. Vanswijgenhoven, E. Wevers, M., Van Der Biest, O.: Surgeon, M., Vanswijgenhoven, E., Wevers, M., & Van der Biest, O. (1999). Transverse cracking and Poisson's ratio reduction in cross-ply carbon fibre-reinforced polymers. Journal of materials science, 34(22), 5513-5517.
5. Van Paepegem, W., De Baere, I., Lamkanfi, E., & Degrieck, J. (2007). Poisson's ratio as a sensitive indicator of (fatigue) damage in fibre-reinforced plastics. Fatigue & Fracture of Engineering Materials & Structures, 30(4), 269-276.
6. De Baere, I., Van Paepegem, W., Degrieck, J.: On the nonlinear evolution of the Poisson's ratio under quasi-static loading for a carbon fabric-reinforced thermoplastic. Part I: Influence of the transverse strain sensor. Polym. Test. **28**(2), 196–203 (2009)
7. Van Paepegem, W., De Baere, I., Lamkanfi, E., Degrieck, J.: Monitoring quasi-static and cyclic fatigue damage in fibre-reinforced plastics by Poisson's ratio evolution. Int. J. Fatigue **32**(1), 184–196 (2010)
8. Pidaparti, R. M., & Vogt, A. (2002). Experimental investigation of Poisson's ratio as a damage parameter for bone fatigue. Journal of biomedical materials research, 59(2), 282-287.
9. Smith, P.A., Wood, J.R.: Poisson's ratio as a damage parameter in the static tensile loading of simple crossply laminates. Composit. Sci. Technol. **38**(1), 85–93 (1990)
10. Gao, F., Boniface, L., Ogin, S.L., Smith, P.A., Greaves, R.P.: Damage accumulation in woven-fabric CFRP laminates under tensile loading: 2. Modelling the effect of damage on macro-mechanical properties. Composit. Sci. Technol. **59**(1), 137–145 (1999)
11. Hill, K.O., Meltz, G.: Fiber Bragg grating technology fundamentals and overview. J. Lightwave Technol. **15**(8), 1263–1276 (1997)

Chapter 4
In Situ High-Rate Mechanical Testing in the Dynamic Transmission Electron Microscope

Michael D. Grapes, Yong Zhang, Melissa K. Santala, Thomas Voisin, Geoffrey H. Campbell, and Timothy P. Weihs

Abstract It is difficult to extract details on deformation mechanisms from conventional high strain rate testing, where microstructural analysis is typically limited to before-and-after comparisons. In situ transmission electron microscopy (TEM) can provide an alternative by allowing direct observation of defect motion during loading, but thus far limitations in the speed of conventional TEM and traditional in situ straining holders have prevented the application of this technique to very high strain rates. We present the latest progress in our efforts to develop such a capability. We have developed a novel TEM specimen holder that uses piezoelectric actuators to pull a specimen in tension at rates up to $10^3\ s^{-1}$. To fit the holder's unique sample geometry we have developed a procedure for fabricating TEM tensile specimens with a consistent, electron-transparent gauge section. These specimens can be fabricated from bulk starting materials, allowing us to retain the materials' original microstructure. The holder is designed to operate in the Dynamic Transmission Electron Microscope (DTEM) at Lawrence Livermore National Lab, which is capable of capturing electron images with exposure times as short as 30 ns.

Keywords: Transmission electron microscopy • In situ testing • Piezoelectrics • Electropolishing • Femtosecond laser machining

4.1 Introduction

The information desired from high-strain-rate mechanical testing can be broadly divided into two main categories: mechanical properties and deformation mechanisms. The former are of obvious importance as they are needed to safely design structures that may be exposed to large dynamic loads. Understanding the deformation mechanisms that drive the mechanical properties is also important because such knowledge enables the development of future materials with tailored dynamic mechanical properties. In terms of mechanical properties, there are now well-established techniques [1] that have been developed that enable the accurate measurement of time-resolved stress and strain over the entire course of a dynamic experiment that may last only fractions of a second. However, resolution of deformation mechanisms has lagged behind this capability. The only practical in situ observation in a conventional high-strain-rate experiment is high-speed videography, which only reveals macroscopic deformation. Analysis of actual deformation mechanisms is limited to comparisons of the sample before and after the deformation has occurred, with characterization typically performed by optical microscopy, scanning electron microscopy, or transmission electron microscopy. This sort of analysis can identify what has *changed* during the experiment, but if the deformation mode changes midway through an experiment it would be missed by such analysis, and it is impossible to learn anything about the *kinetics* of the deformation mechanisms. Looking at it from this perspective it is clear that there is a need for approaches that enable time-resolved, in situ observation of deformation mechanisms in dynamic experiments.

The ability to observe deformation mechanisms in situ already exists for quasi-static loading, in the form of systems that enable loading of an electron-transparent sample inside a transmission electron microscope (TEM). The straining

M.D. Grapes (✉) • T. Voisin • T.P. Weihs
Department of Materials Science and Engineering, Johns Hopkins University, Baltimore, MD 21218, USA
e-mail: mike.grapes@gmail.com

Y. Zhang
Department of Mechanical Engineering, Johns Hopkins University, Baltimore, MD 21218, USA

M.K. Santala • G.H. Campbell
Condensed Matter and Materials Division, Lawrence Livermore National Laboratory, Livermore, CA 94550, USA

B.C. Prorok, L. Starman (eds.), *MEMS and Nanotechnology, Volume 5*,
Conference Proceedings of the Society for Experimental Mechanics Series, DOI 10.1007/978-3-319-22458-9_4

component of these systems has taken a variety of forms, the most popular being a motor-driven system where a low gear ratio enables very precise displacement control [2–6]. This system has the benefit that it can apply a large force and hence is always displacement-controlled, but this feature also makes it unsuitable for high-rate testing. Other approaches have used MEMS sensors and actuators [7, 8] to deform smaller specimens and even capture stress and strain curves in situ. However, none of these devices have been designed for high strain rates.

The present approach is also aimed at in situ TEM investigation, but utilizes piezoelectric actuators to achieve much higher deformation rates than have been possible with previous techniques. While the concept is theoretically sound, its development has presented a variety of challenges that were not encountered with previous approaches. Here, we highlight the basic attributes of the system we have been developing and outline both the problems we have solved and those that are outstanding.

4.2 Holder Design

To accomplish these goals, we have engineered a new holder that uses piezoelectric bending actuators to load a sample in tension. The holder is based on the form factor of the JEOL EM-BST, an older holder providing additional z-space for components at the holder tip while retaining compatibility with the goniometer of the JEOL 2000FX microscope. The holder body is machined out of brass. An overview photograph is shown in Fig. 4.1a. Copper magnet wire runs down the center of the holder to make electrical connections to the piezoelectric actuators at the holder tip. Epoxy is used for the internal vacuum seal. The piezoelectric actuators are bars 25.4 × 3.2 × 0.5 mm (Piezo Systems T220-A3NM-103X). They consist of two sections of piezoceramic material sandwiching a brass shim, and bend in response to an applied voltage. To achieve linear motion at the sample the actuators are mounted in parallel in a clamped–clamped configuration with the sample positioned at their midpoint as shown in Fig. 4.1b. The electrical connections are made such that both actuators pull the sample away from the center when a positive voltage is applied. In the current design both actuators are driven by the same electrical signal resulting in approximately symmetrical displacement, but future designs could employ separate driving signals for each actuator.

Bending actuators were chosen over other styles of piezoelectric actuator (e.g. linear actuators) because of the relatively large forces they can provide (these actuators are rated at 200 mN). However, this value is still very small relative to conventional testing apparatus and necessitates careful design of samples to ensure sufficient stress can be applied to initiate plastic deformation. The sample designs used have in common that, in addition to being electron-transparent in the region of interest, they are also thinned to a lesser extent across the entire center region. These designs are detailed below.

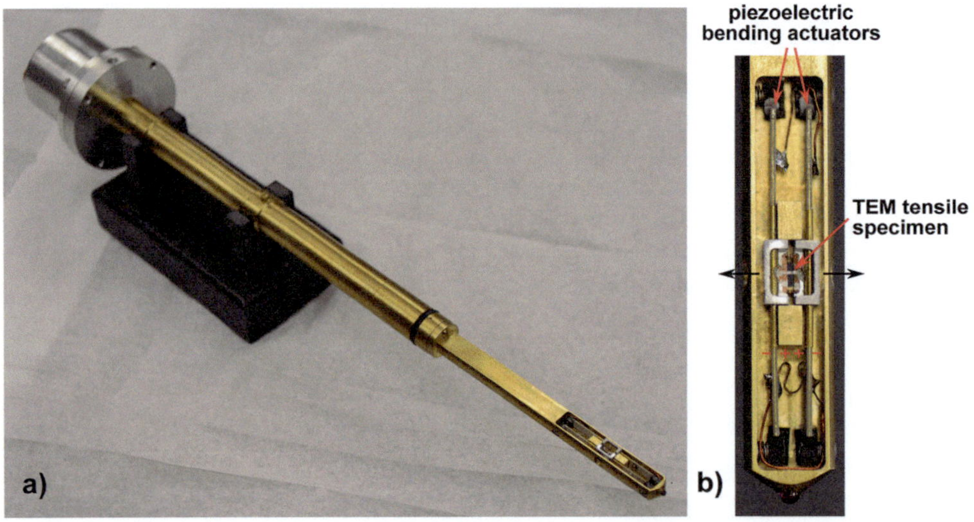

Fig. 4.1 Photographs of (**a**) the entire in situ straining holder and (**b**) details of the holder tip showing how bending actuators are mounted to deliver a tensile load to a specimen

4.3 Sample Preparation

We have explored two approaches for obtaining samples that are both thin enough to deform and perform electron microscopy on and thick enough to be easily handled without damage: standard machining followed by electropolishing and femtosecond laser machining followed by ion milling. These strategies have in common that both allow us to fabricate TEM tensile specimens from bulk samples if required. This is an important capability for our target application.

4.3.1 *Standard Machining and Electropolishing*

Small samples 2.5 mm long and less than 200 μm wide are cut from a 10 μm thick as-rolled foil either by using a razor blade or by electrical-discharge machining. These samples are then electro-polished using a solution of 25 % phosphoric acid, 25 % methanol, and 50 % distilled water for copper and 15 % perchloric acid and 85 % methanol for aluminum. To concentrate the polishing effect only in the center of the sample the samples are mounted on gold aperture grids with a 400 μm hole in diameter. Depending on the positioning of the sample relative to the hole and the polishing time, two styles of thinned samples can be obtained: one where the entire width of the sample is thinned to electron transparency (maximizing the stress that can be achieved but making handling challenging), and one where part of the width remains unpolished, providing more strength for handling at the risk of being too strong to deform in the holder. An outline of this approach is shown in Fig. 4.2.

4.3.2 *Femtosecond Laser Machining and Ion Milling*

Femtosecond laser machining provides the possibility of producing samples with a more well-defined gauge section than can be achieved using electropolishing. Specimens are machined directly from foils of the material of interest. Typical foil thicknesses are 10–25 μm. The current specimen design in use has outer dimensions of 2 × 0.3 mm and a notched gauge section approximately 10 μm wide in the center. The gauge section is thinned to three different levels as shown in Fig. 4.3: one-third ~6 μm thick, one-third ~3 μm thick, and one-third completely open. After laser machining, the in-situ tensile specimen is mounted on a Cu aperture grid for broad-beam ion milling at 4 kV. This reduces the thickness in the 3-μm-thick region to electron transparency. The section that is 6 μm thick is also thinned in this step, but because it starts out thicker it remains thick enough to support handling of the specimen after ion milling.

4.3.3 *Mounting Specimens for Testing*

Regardless of preparation strategy, the samples must be mounted securely on the sample holder for experimentation. The strategy for doing this has changed several times in the course of development. In the current approach the samples are

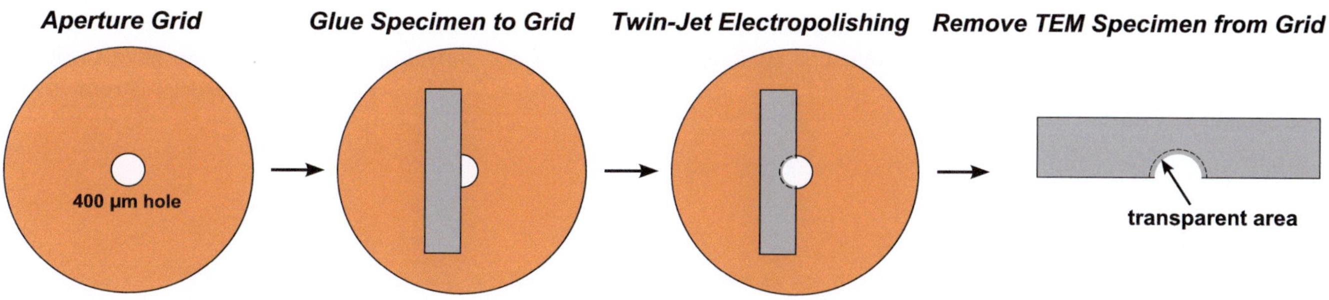

Fig. 4.2 Basic processing sequence for using electropolishing to produce a TEM tensile specimen, in this case of the half-polished variety

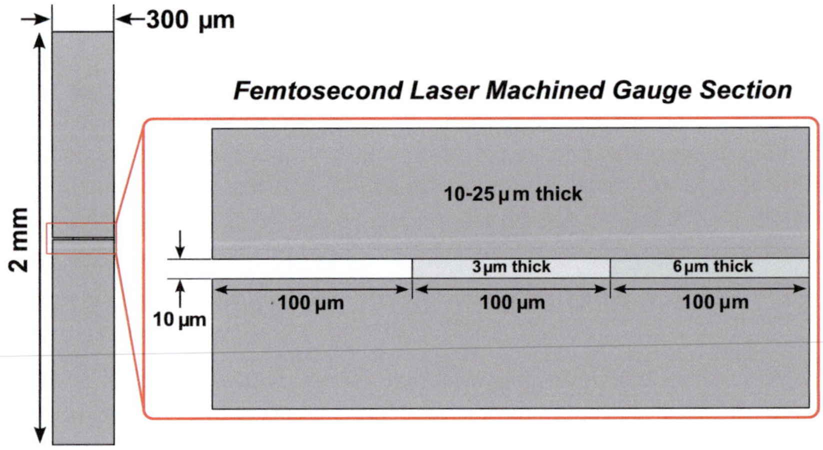

Fig. 4.3 Proposed design for samples produced by femtosecond laser machining

mounted on a copper frame using silver paint and super glue. The frame provides support for handling the sample and can be easily installed in the holder. Silver paint is used to ensure electrical conductivity from the sample to ground. Since the actuators are not strong enough to deform a copper frame as well as the sample, the frame actually consists of two halves that are initially joined using super glue. This glue is removed to separate the frames just before experimentation. A sample loaded using this approach can be seen in the center of Fig. 4.1b.

4.3.4 Operating the Holder

For the majority of each experiment the piezoelectric actuators are operated in a force-limited mode. In this mode, the force applied is roughly proportional to the applied voltage while the displacement is determined by the deformation of the sample. To some extent the strain rate achieved can be varied by altering the rate at which force is applied, but this is an indirect control at best, and notably different from the situation in conventional high strain-rate mechanical testing like the split-Hopkinson bar where the system is over-driven such that it is always displacement-controlled.

To deal with this shortcoming, we have conducted extensive characterization of actuator performance in response to various voltage inputs using both reflectance-based displacement sensors and high-speed videography. Taken as a whole, these experiments allow us to design the voltage waveform used to achieve an approximately constant, predictable strain rate.

4.4 Dynamic Transmission Electron Microscope

The dynamic TEM is a modified JEOL 2000FX transmission electron microscope at Lawrence Livermore National Laboratory in which the conventional electron gun has been replaced with a laser-driven Ta photocathode. Electrons are generated by bombarding the photocathode with a high-intensity ultraviolet laser. This approach supplies sufficient electron exposure for a single image in as little as 30 nanoseconds. By applying a train of laser pulses to the photocathode, a sequence of time-resolved electron images can be obtained spanning 50–100 μs. Since conventional TEM does not have sufficient time-resolution to capture events at high strain rates, the dynamic TEM (DTEM) is the ultimate target instrument for this system.

4.5 Preliminary Results

Because of the complexity of the combined straining holder + dynamic TEM system, no time-resolved electron images during rapid loading have been obtained as of this writing. In preparation for this task, preliminary tests have involved (1) testing the loading system at high strain rates outside the TEM, and (2) testing the loading system at quasi-static strain rates inside the TEM. Integration with the DTEM represents the combination of these two precursors, but presents significant new challenges in terms of timing and imaging that we are still in the process of solving.

For ex situ high-rate tests, we place the holder under a conventional light microscope with 1.4×–18.0× magnification and record videos of straining experiments using a high speed camera. The resulting videos can be used to identify how the samples deform, where most of the deformation occurs, and to approximate strain rates. The latter is accomplished by digital image correlation (DIC) analysis of the natural speckle pattern on polished specimens. In Fig. 4.4 we show an example of the results obtained using this technique on a sample prepared using the electropolishing approach outlined above.

In situ quasi-static tests have been performed at Lehigh University in Bethlehem, PA, where there is a JEOL 2000FX TEM compatible with the holder. In these tests, we first identified a region where deformation was likely to be concentrated (a crack-tip) and then incrementally increased the voltage applied to the piezoelectric actuators to so that a force was applied that tended to pull the crack open. Again, the images shown were taken from a sample prepared by electropolishing. Sequential images of the crack as shown in Fig. 4.5 illustrate our ability to observe large-scale deformation inside the TEM. However, they do not reveal the underlying deformation mechanism (dislocations) adequately, an important capability for future experiments. In fact, these tests revealed several issues with both holder design and sample preparation that were preventing us from obtaining high-quality images of dislocations. These issues have been addressed and new results are forthcoming.

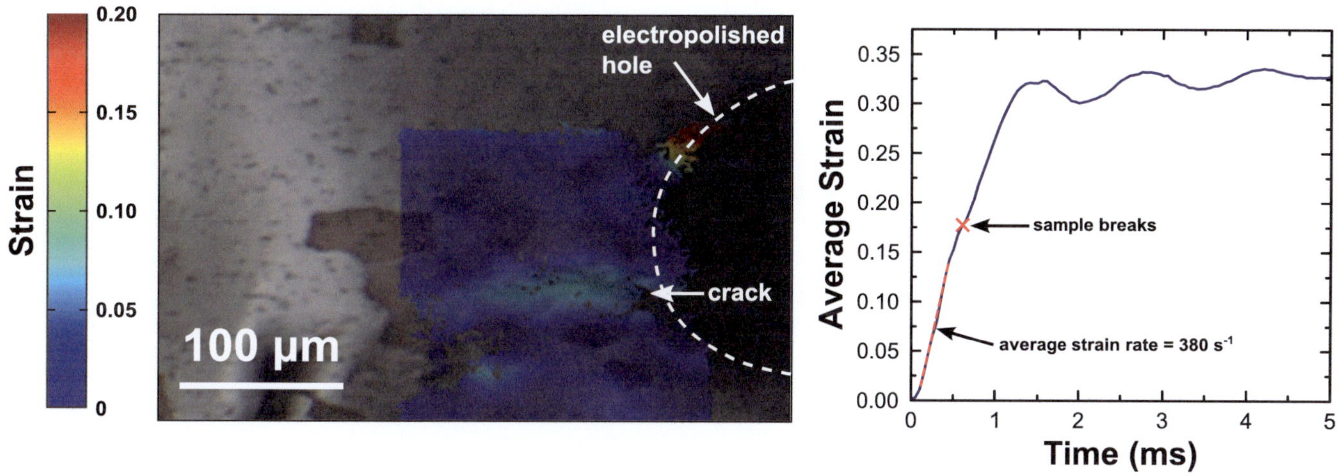

Fig. 4.4 One frame from a high-speed video taken of an electropolished sample being deformed inside the holder. A DIC strain map is overlaid over the region where deformation is occurring. At *right*, DIC results as a function of time are used to estimate the strain rate in this experiment

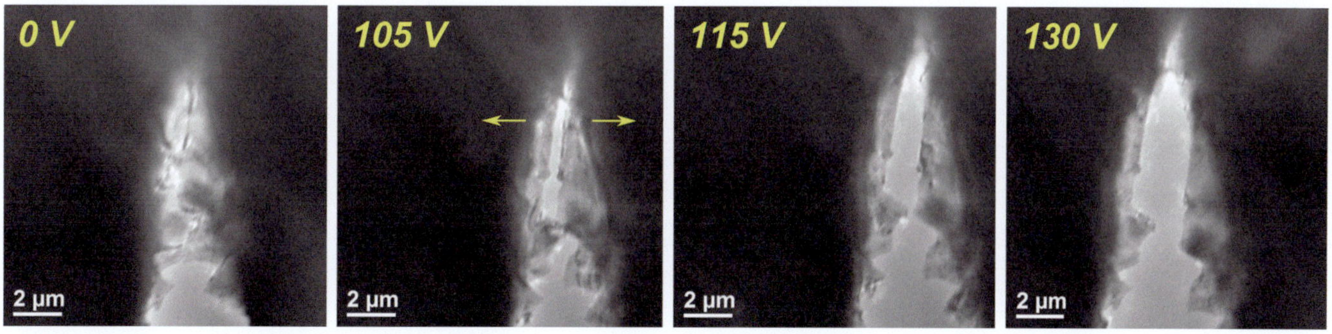

Fig. 4.5 Sequence of bright-field electron images captured during in situ quasi-static loading, showing a crack opening as the piezo voltage (force applied) is increased

4.6 Conclusion

We have developed a system capable of deforming metals at high strain rates inside a transmission electron microscope. When combined with a time-resolved electron microscope like the dynamic TEM at Lawrence Livermore National Laboratory, the system is expected to deliver time-resolved images of the underlying deformation mechanisms operating at strain rates as high as 10^3 s^{-1}. A variety of metals can be tested using this system, limited only by sample preparation protocols and geometry restrictions based on the finite force available from the piezoelectric actuators. We anticipate our first time-resolved in situ results by the end of 2015, and possibly in time for this meeting.

Acknowledgements Thanks to Rob Keyse and the Center for Advanced Materials and Nanotechnology at Lehigh University for TEM facilities. Funding was provided by the U.S. Army through the Materials in Extreme Dynamic Environments Cooperative Research Agreement. Development of the femtosecond laser machining capability is conducted in collaboration with Brian Schuster, Jonathan Ligda, and Nicholas Lorenzo of the U.S. Army Research Laboratory.

References

1. Field, J.E., Walley, S.M., Proud, W.G., Goldrein, H.T., et al.: Review of experimental techniques for high rate deformation and shock studies. Int. J. Impact Eng. **30**, 725–775 (2004)
2. Kröger, A., Dziaszyk, S., Frenzel, J., Somsen, C., et al.: Direct transmission electron microscopy observations of martensitic transformations in Ni-rich NiTi single crystals during in situ cooling and straining. Mater. Sci. Eng. A **481–482**, 452–456 (2008)
3. Caillard, D.: TEM in situ straining experiments in Fe at low temperature. Philos. Mag. Lett. **89**, 517–526 (2009)
4. Legros, M., Gianola, D.S., Hemker, K.J.: In situ TEM observations of fast grain-boundary motion in stressed nanocrystalline aluminum films. Acta Mater. **56**, 3380–3393 (2008)
5. Robach, J.S., Robertson, I.M., Wirth, B.D., Arsenlis, A.: In-situ transmission electron microscopy observations and molecular dynamics simulations of dislocation-defect interactions in ion-irradiated copper. Philos. Mag. **83**, 955–967 (2003)
6. Hugo, R.C., Kung, H., Weertman, J.R., Mitra, R., et al.: In-situ TEM tensile testing of DC magnetron sputtered and pulsed laser deposited Ni thin films. Acta Mater. **51**, 1937–1943 (2003)
7. Zhu, Y., Espinosa, H.D.: An electromechanical material testing system for in situ electron microscopy and applications. Proc. Natl. Acad. Sci. U. S. A. **102**, 14503–8 (2005)
8. Haque, M.A., Saif, M.T.A.: In situ tensile testing of nanoscale freestanding thin films inside a transmission electron microscope. J. Mater. Res. **20**, 1769–1777 (2011)

Chapter 5
In Situ TEM Observation of Twinning, Detwinning and Retwinning in Quartz

E. Tochigi, E. Zepeda, H.-R. Wenk, and A.M. Minor

Abstract Quartz nanopillars were fabricated by focused ion beam technique from a natural crystal. The nanopillars were deformed by in situ compression test in a transmission electron microscope. During loading and unloading operations, nucleation, extension and contraction of Dauphiné twins were observed in a pillar. The twinning phenomena appear to include a memory effect, where the same twin can reappear upon successive loading and unloading events. We discuss the twinning phenomena with mechanical data taken during the compression test.

Keywords: Quartz • Twinning • In situ • TEM • Compression

5.1 Introduction

Quartz has a 180° rotation twin about the *c*-axis. This is called Dauphiné twin and is mechanically induced with a small distortion of bonding in the crystal structure [1, 2]. This behavior is a result of the high elastic anisotropy in quartz, where the direction close to perpendicular to the negative rhomb {−1011} is almost twice as stiff as one perpendicular to the positive rhomb {10-11} [3]. Compression perpendicular to the negative rhomb flips the orientation to the positive rhomb with minimal macroscopic strain. However, little is known about the twinning mechanism and the necessary stresses to induce the twinning. In this report, we outline dynamic observations and measurement of twinning behavior of quartz nanopillars using a quantitative mechanical deformation stage inside a transmission electron microscope (TEM). A more detailed discussion on deformation of quartz by twinning as well as by dislocation slip and by amorphization on the basis of nanopillar compression study is given in [4].

5.2 Experimental Procedure

A natural quartz crystal was cut to a small chip. The orientation of the sample was measured by electron back scatter diffraction (EBSD) to orient the crystal for compression test. The sample was mechanically ground to obtain a thickness of about 30 μm. The sample was mounted on a half-moon shaped supporting mesh. Amorphous carbon was deposited on both the surfaces of the sample to minimize charging problems during milling by focused ion beam (FIB) technique. Nanopillars were fabricated by FIB operated at 30 kV with a Ga^+ ion source (Strata DB 235, FEI, USA). The long axis of the pillar is approximately perpendicular to the {−1011} plane (Fig. 5.1), which is close to the stiffest direction of quartz. One side of the sample was coated by thin amorphous carbon to prevent charging during TEM observation.

E. Tochigi (✉)
Institute of Engineering Innovation, The University of Tokyo, Tokyo, Japan
e-mail: tochigi@sigma.t.u-tokyo.ac.jp

E. Zepeda • H.-R. Wenk
Department of Earth & Planetary Science, University of California, Berkeley, CA 94720, USA

A.M. Minor
Department of Materials Science & Engineering, University of California, Berkeley, CA 94720, USA

National Center for Electron Microscopy, Molecular Foundry, Lawrence Berkeley National Laboratory, Berkeley, CA 94720, USA

B.C. Prorok, L. Starman (eds.), *MEMS and Nanotechnology, Volume 5*,
Conference Proceedings of the Society for Experimental Mechanics Series, DOI 10.1007/978-3-319-22458-9_5

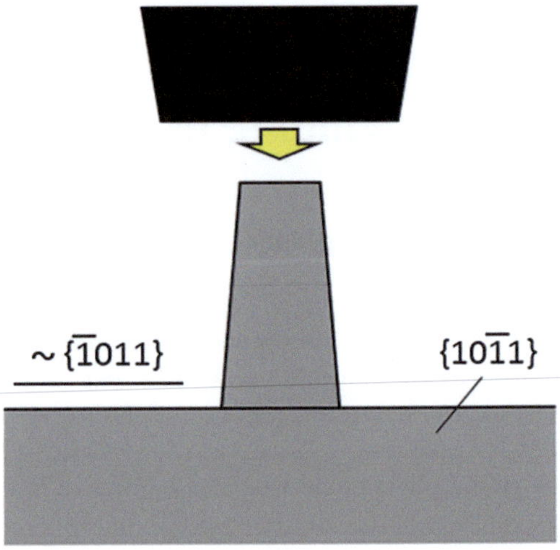

Fig. 5.1 Schematic illustration showing crystal orientation for in situ compression test. The long axis of the pillar is approximately perpendicular to the {−1011} plane, which is close to the stiffest direction of quartz

In situ compression experiment was performed on nanopillars by using an indentation sample holder with a diamond flat punch (PI-95, Hysitron, USA) and a TEM operated at 300 kV (JEM-3010, JEOL, Japan). The compression was executed under displacement control of the loading rate at 3 nm/s. During compression experiment, structural changes of the pillars were recorded as a movie at 10 frames per second by a CCD camera for TEM (SC200D, Gatan, USA).

5.3 Results and Discussion

Figure 5.2 illustrates twinning of a quartz pillar by in situ TEM compression testing. Figure 5.2a shows a TEM image of the pillar before compression testing taken under a g-12-11 two-beam condition. The diameter of the pillar is ~470 nm at the top and the taper angle is ~4°. No obvious defect contrast can be seen. Still images captured from the video recorded during the compression testing are shown in Figs. 5.2b–g. Figure 5.2h shows a TEM image of the pillar after compression testing. A plot of engineering stress versus displacement of the test is shown in Fig. 5.3. Labels (b)–(g) on the plot indicate the position of the captured images in Fig. 5.2. A significant strain burst can be seen at point (b). It corresponds to the appearance of a dark contrast feature of about 400 nm in length (arrow 1 in Fig. 5.2b) and relatively weak curved contrasts over the pillar, which would correspond to dislocations. The dark contrast grows upon further loading (Fig. 5.2c) and evolves into a knife-shaped feature in Fig. 5.2d. An additional band of dark contrast (arrow 2 in Fig. 5.2d) leads to a second knife-shaped feature as shown in Fig. 5.2e, partially overlapping the first one. Upon unloading (Figs. 5.2e–h), both features progressively disappear. At first, the second feature shrinks to the position pointed out by the arrow 3 in Fig. 5.2f, and then the first feature shrinks towards the left-hand side of the pillar (Fig. 5.2g) until they both completely disappear (Fig. 5.2h).

The strain burst associated with the point (b) in Fig. 5.3 indicates the formation and movement of the initial defect nucleus. At this moment, the pillar itself demonstrates almost no bending; namely, the crystal orientation should be the same. It is unlikely that this contrast corresponds to some pre-existing defects visualized by crystal orientation change. In addition, we confirmed a similar evolution of knife-shaped contrast in different pillars induced by in situ compression testing. Indeed, the characteristic deformation feature observed during nanocompression tests was a planar, knife-shaped contrast with a sharp corner. In addition, the two knife-shaped contrast features shown in Fig. 5.2 separately nucleated, propagated, and overlapped. Thus, it is reasonable to consider that the contrast changes observed in Fig. 5.2 correspond to the evolution of two individual defects. Further, the disappearance of the features upon unloading suggests that the features are not due to dislocation plasticity. Instead, we attribute these features to Dauphiné twins.

After the strain burst (b in Fig. 5.3), the slope of the engineering-stress plot decreases from 29.1 MPa/nm to 22.4 MPa/nm and is almost constant (constant slope from c to d in Fig. 5.3). The in situ observation shows that this decrease of slope is a result of the propagation of the Dauphiné twins and the dislocations. The contribution of plasticity due to the dislocations would be small because only a few dislocations were generated. Thus, the decrease of slope indicates that the deformation resistance of the pillar decreased as mediated by Dauphiné twinning. This is consistent with the stiffest direction flipping to the softest direction by Dauphiné twinning, as expected from the orientation relationship in our experiment.

The local minimum stresses at the strain burst were about 970 MPa, and the diameter of the pillar at the nucleation point of the first twin was ~530 nm, which is ~13 % larger than that of the pillar top. Thus, the nominal stress for the twinning is

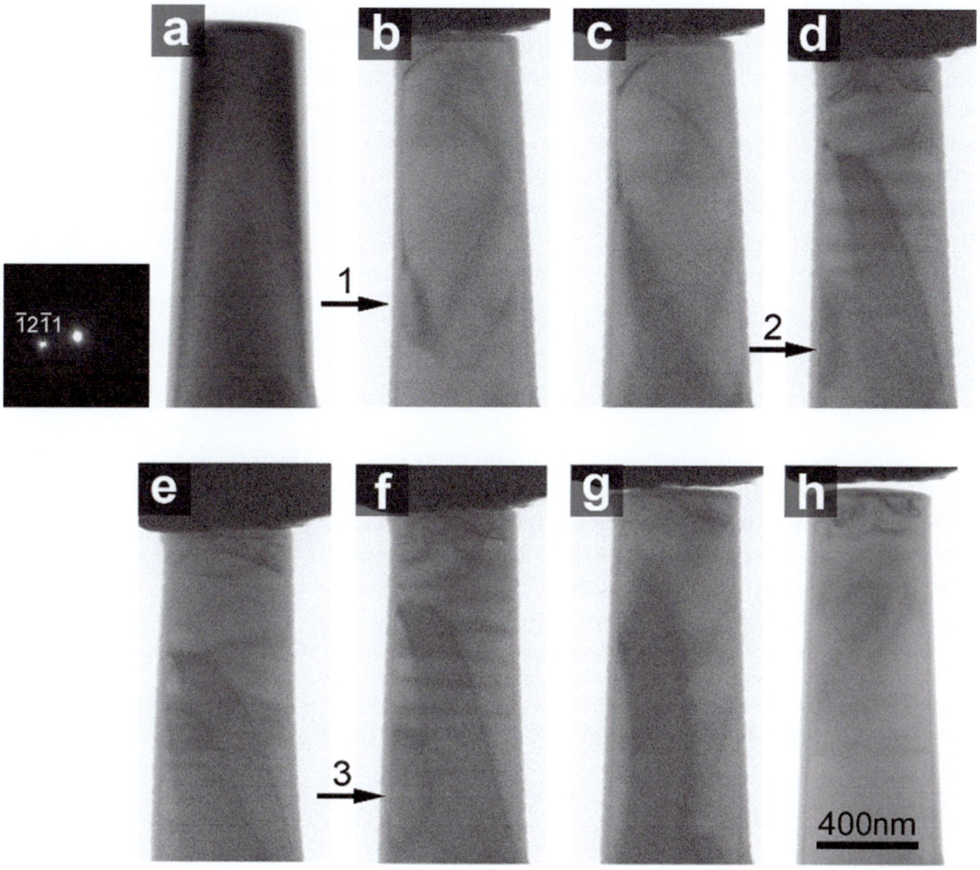

Fig. 5.2 Deformation of a quartz pillar by twinning. (**a**) A bright-field TEM image using g-12-11 before compression and the corresponding diffraction pattern aligned to the image, where the {−12-11} plane is almost parallel to the incident direction and the {−1011} plane is a few degrees off to the incident direction or to the plane of the pillar top. (**b–e**), Still images captured from a movie recorded upon loading. Two different twins nucleated from the points indicated by arrows 1 and 2 in **b** and **d**. (**e–g**), Still images captured from a movie recorded upon unloading. The second twin shrank into arrow 3 in **f**. The first twin also shrank into the *left-hand side* of the pillar. (**h**) A TEM image after compression. The characteristic twin features are disappeared

Fig. 5.3 A plot of engineering stress versus displacement taken through the test. The labels b–g correspond to the captured image with the same notations in Fig. 5.2

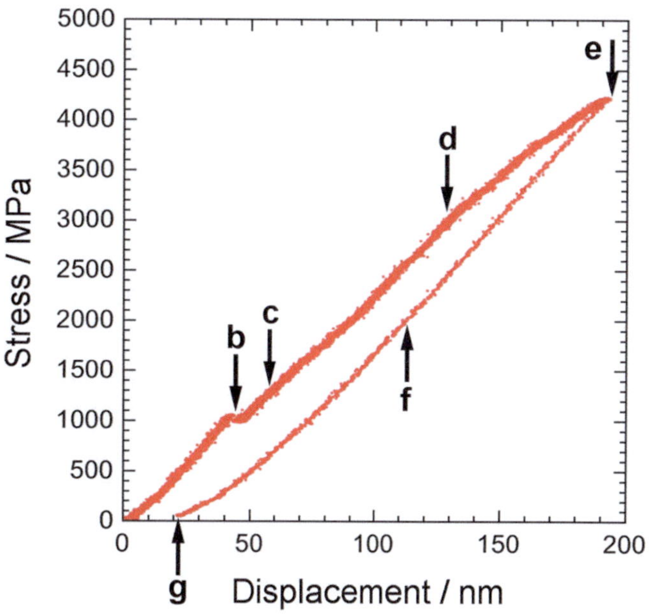

calculated to be 760 MPa. It was reported that Dauphiné twins are formed by a nominal average stress of about 200 MPa in polycrystalline quartz [5], but the local stress for twinning has never before been reported. In addition, the twins continued to grow until the nominal stress reached 3320 MPa, and then they disappeared with unload. This implies that the Dauphiné twins created here are essentially unstable without an applied stress.

We recompressed the same pillar subsequently to the compression shown in Fig. 5.2. The deformation behavior during the second compression was quite similar [4], indicating that the twin propagation is remarkably repeatable, in effect a retwinning of the same twin due to a memory-effect. Interestingly, the plot of the second compression had no strain burst and its slope was almost the same to that of the first compression after the strain burst, suggesting that the twin nucleation processes differs between the first and the second compression.

5.4 Conclusion

Dauphiné twinning in quartz crystals was investigated by in situ nanocompression experiments in a TEM. Our results clearly showed that dynamic behavior of twinning, detwinning and retwining in quartz. With twin nucleation the strain burst was shown and deformation resistivity significantly decreased. However, in the subsequent compression test, no strain burst was seen and the deformation resistivity remained decreased, although the same twins nucleated and propagated. These results imply that quartz crystals can contain a memory of their Dauphiné twinning history.

Acknowledgements The authors acknowledge support of the National Center for Electron Microscopy, Lawrence Berkeley National Laboratory, which is supported by the U.S. Department of Energy under Contact # DE-AC02-05CH11231. ET was supported by JSPS Postdoctoral Fellowships for Research Abroad and “Nanotechnology Platform” (project No.12024046) of the Ministry of Education, Culture, Sports, Science and Technology (MEXT), Japan. HRW acknowledges support from National Science Foundation (EAR 1343908) and DOE-BES (DE-FG02-05ER15637).

References

1. Schubnikov, A., Zinserling, K.: Ueber die Schlag- und Druckfiguren und ueber die mechanischen Quarzzwillinge. Z Kristallogr **83**, 243–264 (1932)
2. Hahn, T., Klapper, H.: Twinning of crystals. In: Authier, A. (ed.) International tables for crystallography, vol. D, pp. 393–448. Kluwer Academic Publishers, Dordrecht (2003)
3. McSkimin, H.J., Andreatch, P.J., Thurston, R.N.: Elastic moduli of quartz versus hydrostatic pressure at 25° and −195.8 °C. J. Appl. Phys. **36**, 1624–1632 (1965)
4. Tochigi, E., Zepeda-alarcon, E., Wenk, H.R., Minor, A.M.: In situ TEM observations of plastic deformation in quartz crystals. Phys. Chem. Miner. **41**, 757–765 (2014)
5. Wenk, H.R., Bortolotti, M., Barton, N., Oliver, E., Brown, D.: Dauphiné twinning and texture memory in polycrystalline quartz. Part 2: In situ neutron diffraction compression experiments. Phys. Chem. Miner. **34**, 599–607 (2007)

Chapter 6
Nano to Macro: Mechanical Evaluation of Macroscopically Long Individual Nanofibers

D. Papkov[$], K. Maleckis[$], Y. Zou, M. Andalib, A. Goponenko, and Y. Dzenis

Abstract Nanomaterials such as carbon nanotubes and graphene attract considerable attention due to their extraordinary mechanical and other properties. However, discontinuous nature of these carbon allotropes prevents easy transfer of their mechanical properties to the macro scale. Continuous nanofibers represent an emerging class of nanomaterials with critical advantages for structural and functional applications. However, their mechanical testing to date has been largely conducted using micrometer-long specimens in AFM-type or MEMS devices. In addition, most published reports did not test nanofibers through failure. As a result, information relevant to potential macroscopic structural applications of nanofibers is currently very limited. Here, we will present and discuss a recently developed, comprehensive mechanical evaluation protocol spanning controlled nanomanufacturing, handling, and mounting of long individual nanofiber specimens, as well as analysis of their large-deformation behavior through failure, and data reduction. The protocol will be demonstrated on several types of synthetic and biological nanofibers, including nanofibers exhibiting unique simultaneously ultrahigh elastic modulus, strength, and deformation to failure, resulting in superhigh toughness. The developed protocol will be instrumental for further optimization of mechanical properties of continuous nanofibers.

Keywords: Continuous nanofibers • Electrospinning • Mechanical testing • Size effects • Diameter control

6.1 Introduction

Since the development of cast iron technologies in 1620s and steels in 1850s, metals dominated the structural materials' applications through the first decades of the twentieth century. However, since 1960s the development of new metal alloys is slow, while polymer and composite industries had expanded rapidly [1]. Combining several materials to create a composite material has been known throughout human history, however modern era of composites can be said to begin approximately at the time of WW II [2]. The properties of advanced composites are driven by the properties of advanced fibers used in them. Several key developments in 1960s spurred the development of this new generation of materials and allowed them to revolutionize the structural material markets and to make composites ubiquitous in our daily lives today [3]. While industrial R&D continues to this day, mechanical properties of advanced fibers largely plateaued, and most efforts are directed to cost reduction and incremental improvements [4]. For example, no major breakthroughs in mechanical properties of these fibers were reported since late 1980s [5].

Nanomaterials and nanotechnology provide the potential for the next step in material and processing development. However, many problems remain with the realization of this potential. Ever since the discovery of carbon nanotubes and more recently graphene and graphene oxide, and subsequent discovery of their outstanding mechanical properties these materials were the subject of intensive research efforts and financial investment. Unfortunately the mechanical properties had fallen far short of predictions and multiple fundamental problems remain, limiting the structural applications of these carbon nano-allotropes [6, 7]. The problems are well documented and, at their core, stem from the discontinuous nature of these nanoparticles [8].

Continuous nanofibers (NFs) are an emerging class of materials offering the potential of unique property combinations, ways to overcome the problems associated with the use of discontinuous nanoparticles, and to alleviate potential health hazards. There are multiple techniques for manufacturing NFs. Electrospinning is a popular, inexpensive top-down process

[$]The two authors contributed equally to this research.

D. Papkov • K. Maleckis • Y. Zou • M. Andalib • A. Goponenko • Y. Dzenis (✉)
Department of Mechanical and Materials Engineering, University of Nebraska—Lincoln,
W315 Nebraska Hall, Lincoln, NE 68588-0526, USA
e-mail: ydzenis@unl.edu

B.C. Prorok, L. Starman (eds.), *MEMS and Nanotechnology, Volume 5*,
Conference Proceedings of the Society for Experimental Mechanics Series, DOI 10.1007/978-3-319-22458-9_6

with multiple controls that offers high flexibility to produce continuous nanofibers from a wide range of materials and architectures. The process had been widely utilized and studied since mid-1990s, due to its potential use in a variety of applications. During electrospinning a solution or a melt of a polymer is jetted at high electric field from spinneret to collector once the electric field overcomes the surface tension of the liquid. The jet then undergoes large elongation under high forces, several instabilities due to electrostatic repulsion, solidifies or dries (usually in-flight) and is deposited on the collector.

A large number of polymer systems had been electrospun and NFs with diameters from several tens of nanometers up to several micrometers were obtained. Despite the apparent manufacturing simplicity, the process is extremely complex, involves multiple physical phenomena, and is not fully understood. While the flexibility of the process allows for a large number of control parameters, it also introduces significant complexity in determining structure/property relations. Some attempts of modeling the process and the consequent properties of NFs were done over the years, but they so far fail to capture the full complexity of the process, and many of the aspects still lack empirical data needed for a proper evaluation [9].

6.2 Methods of Mechanical Testing of 1D Nanomaterials

Most of the early mechanical testing done on electrospun NFs, used random and partially oriented mats or bundles of NFs [9, 10]. The attractiveness of this approach is the fact that regular mechanical testing methods used for material characterization can be applied. In addition, this approach allows direct translation of the properties to applications, and is directly relevant for specific applications where such NF assemblies are used, such as tissue engineering [10]. While some interesting properties were reported for mechanical testing of these assemblies, such as increased modulus compared to the bulk material, and significant improvement in properties with the increase of preferred NF orientation within the assembly [9, 10], these results remain inconsistent across different polymer systems [9], and as a result of these tests, electrospun materials were generally considered weak [8].

One of the main reasons for poor mechanical performance is wide diameter distribution of NFs within the assembly (see next section). The properties of the assembly depend both on the macroscopic parameters such as contact density, NF alignment, porosity etc., but also on the difference in properties of individual NFs within the assembly. Correlation of any one parameter to mechanical performance and determination of structure/property relations in these cases is next to impossible, because the different parameters are extremely hard to disentangle without understanding the property relations on the individual NF level first [9].

Evaluation of mechanical properties and structure of individual electrospun NFs is not easy, and it is still a very active research field. Most of the approaches rely on specialized AFM techniques or MEMS devices [9]. Already in 1996, Jaeger et al. [11] used AFM to examine packing of polymer chains on the surface of the electrospun NFs. Development of AFM based techniques to examine mechanical properties of carbon nanotubes in early years of the twenty-first century [12, 13] allowed these methods to be later applied to individual NFs. In recent years, several studies reported on strong size effects in mechanical properties of individual NFs (a partial summary can be seen in ref. [14]).

The mechanical testing techniques on the individual NF level can be broadly broken down into several categories [9, 15]. The easiest to implement are *nanoindentation techniques* [16–19] (see schematic in Fig. 6.1a), however this techniques only provides a limited information on the properties of electrospun NFs, mostly confined to surface modulus.

Bending such as used by [20–22] (see schematic in Fig. 6.1b) or a more specialized resonance technique [23] rely on elastic behavior of NFs in the examined region and also provides information limited to the modulus of the NFs. While it is possible to adapt a bending technique to testing through failure, such a technique would still have a problem with examining large deformations, and extraction of different parameters is not straightforward with the onset of plastic deformation.

A more comprehensive examination of NF behavior can be obtained through *tensile tests* of individual NFs [24–28] (see schematic examples in Fig. 6.1c–e). A similar technique was also recently used in collaboration between our group and the Espinosa group at Northwestern University [29]. Most of these tests rely on a specialized MEMS devices. When these are combined with an SEM or TEM stage (see *e.g.* [29]), they can provide information on NF structure and failure mechanisms. While these techniques provide a more comprehensive approach to the examination of the mechanical properties of individual NFs, some aspects (such as systematic examination of strain to failure) remain underreported in the literature. In addition, there are several issues having a potentially large effect on the results. In the above techniques, small length

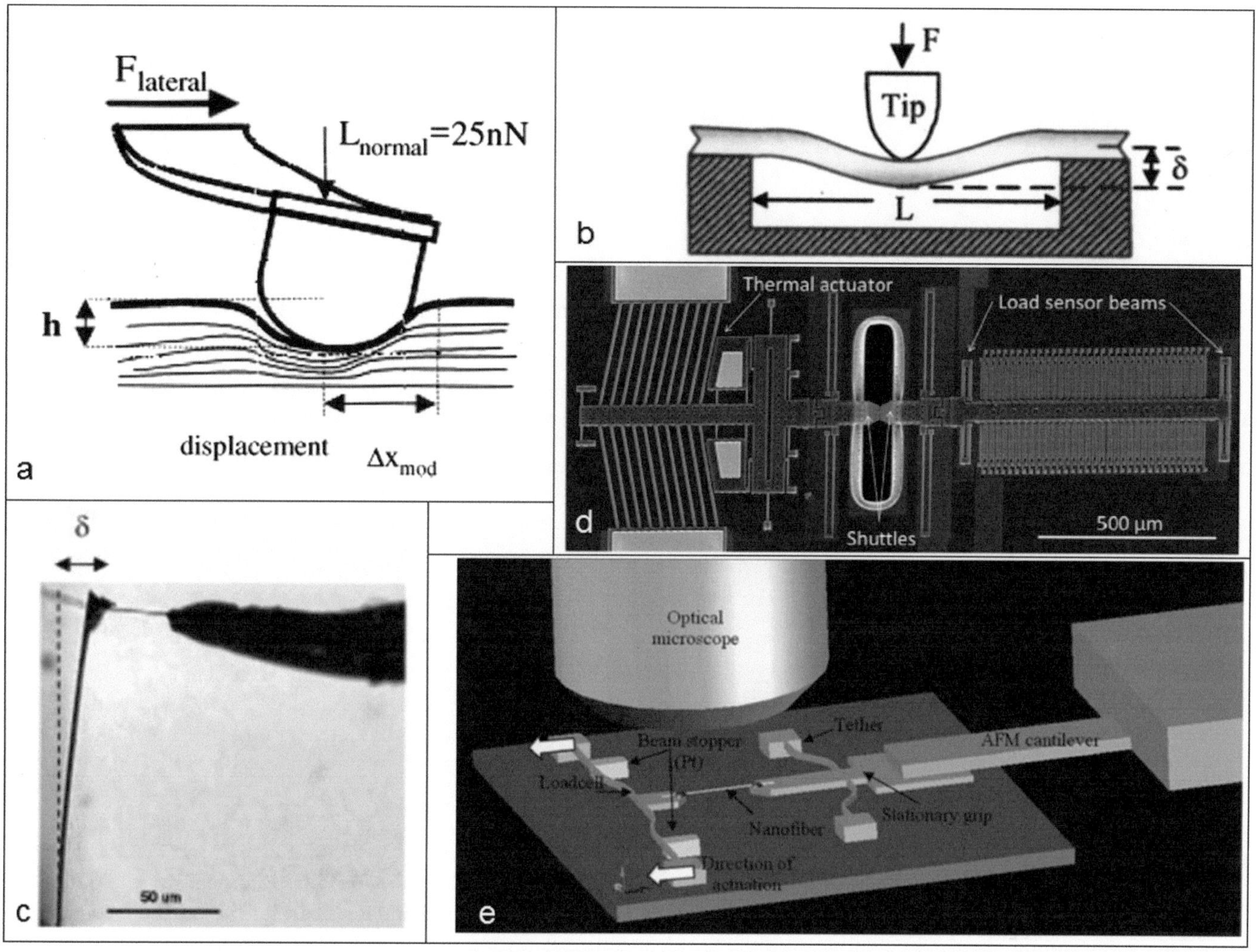

Fig. 6.1 Different mechanical testing techniques in the literature. (**a**) Nanoindentation (adopted from [16]); (**b**) bending (adopted from [21]); Tensile testing using (**c**) AFM tip (adopted from [25]), and MEMS devices (**d**) adopted from [29], (**e**) adopted from [27]

samples are used (several tens of micrometers compared to several centimeters used for commercial fibers). As a result, gripping effects (such as slipping of the fibers) become potentially more influential. In addition, in-situ techniques are in some cases poorly suited for examinations of large deformations due to simple space limitations.

Beyond these effects that can be overcome with appropriate sample preparation and choice of testing system, length effects in mechanical properties constitute a significant concern. Fracture mechanics and critical flaw analysis [30, 31] predict significant increases in material strength as the length of the sample decreases by approximately two orders of magnitude. While the applicability of the Weibull distribution analysis and other scaling laws might be limited [32] (and of relevance to brittle rather than elasto-plastic materials), the question of length effects in the reported results remains and puts a question mark on an effective comparison between the different studies as well as complicating the upscaling of these results to the macro scale. Length effects are not to be confused with the effects of NF diameter, since the latter are more likely to be structure dependent and are in-built through the distribution of NF diameters within NF assemblies.

As a result, a comprehensive testing protocol that can examine macroscopically long individual NFs through failure (including in cases of large deformations) is of great interest and importance. In addition, examination of mechanical performance of NFs as a function of their diameter is also paramount if any successful attempt to translate the properties of individual NFs to NF assembly properties is to be made in the future.

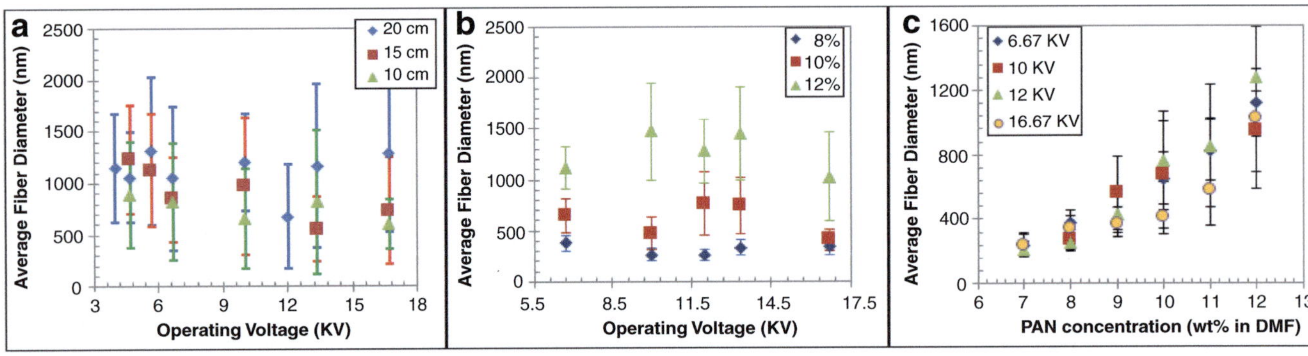

Fig. 6.2 Effect of process parameters during electrospinning on PAN NF diameter distributions. (**a**) Changes in average fiber diameters across the range of operating voltage for different spinneret-collector distances (all tests were done with 10 wt% PAN concentration in DMF); (**b**) changes in average fiber diameters across the range of operating voltage for different polymer concentrations (all tests were done for a 20 cm spinneret-collector distance); (**c**) changes in average fiber diameters across the range of PAN concentrations for different operating voltages (all tests were done for a 20 cm spinneret-collector distance). In all cases the error bars represent the standard deviation of the fiber diameter distribution

6.3 Control of Nanofiber Diameter

Control of NF morphology and diameter distribution is paramount and has a profound impact on the NF physical, chemical electronic, and mechanical properties. As was discussed above, previous studies showed significant size effects in mechanical properties of NFs, underscoring the importance of this control. Electrospinning is a complex, multi-physics process with multiple intertwined parameters impacting the eventual morphology, diameter, and properties of NFs (for a more complete discussion of the different parameters in determining morphology, diameter, and properties of NFs see [8]).

Effects of electric field and polymer concentration on NF diameters were examined for polyacrylontrile (PAN) electrospun from dimethylformamide (DMF). A range of spinneret/collector distances, operating voltage, and PAN concentration by weight were examined (see Fig. 6.2). Operating voltage (Fig. 6.2a–b) did not have a significant effect on the average diameter (especially within the context of the relatively wide diameter distributions) in the range of electrospinning distances and PAN concentrations studied. The breadth of the diameter distribution (as indicated by the error bars representing standard deviations within the distributions) also did not noticeably change. Generally, smaller spinning distances produced slightly thinner nanofilaments (on average). When examining the effect of PAN concentration, it was found that uniform NFs could be obtained in the range of 9–12 wt% concentrations of PAN in DMF. Some beading was observed for 8 wt% concentration, which became significantly worse for the 7 wt% concentration. PAN concentration constituted the biggest effect on average NF diameter and diameter distribution width, with both parameters sharply increasing with increase in polymer concentration (Fig. 6.2c).

It should be noted that the wide diameter distributions observed in experiments, coupled with the sharp size effects reported for mechanical properties of individual NFs, also explain the relatively poor properties observed in mechanical testing of NF assemblies even when the quality of such assemblies was high.

6.4 The Protocol for Mechanical Testing and Data Reduction

A protocol for sample preparation and mechanical testing through failure of individual electrospun NFs was developed. This protocol was applied to multiple synthetic and natural polymer systems, as well as composite, and carbon nanofibers. In this paper we describe the protocol and the results obtained for PAN NFs. In a later part of the paper we also describe a modification of this protocol already applied for ultrathin DNA nanofibers (see a more detailed description in [8, 14, 33]).

Individual PAN NFs in a wide range of diameters (controlled through process parameters described in the previous section) were electrospun on a window collector, and picked up by a "fork-like" collector (see Fig. 6.3a). To avoid possible radiation damage on the mechanically tested part, each NF was split into two parts (see Fig. 6.3b), one of which was then examined in an SEM to determine the NF diameter. Nanomechanics testing system from MTS was used for mechanical testing of the individual NFs. To avoid damage to the fibers during gripping and template cutting, the second part of the fiber

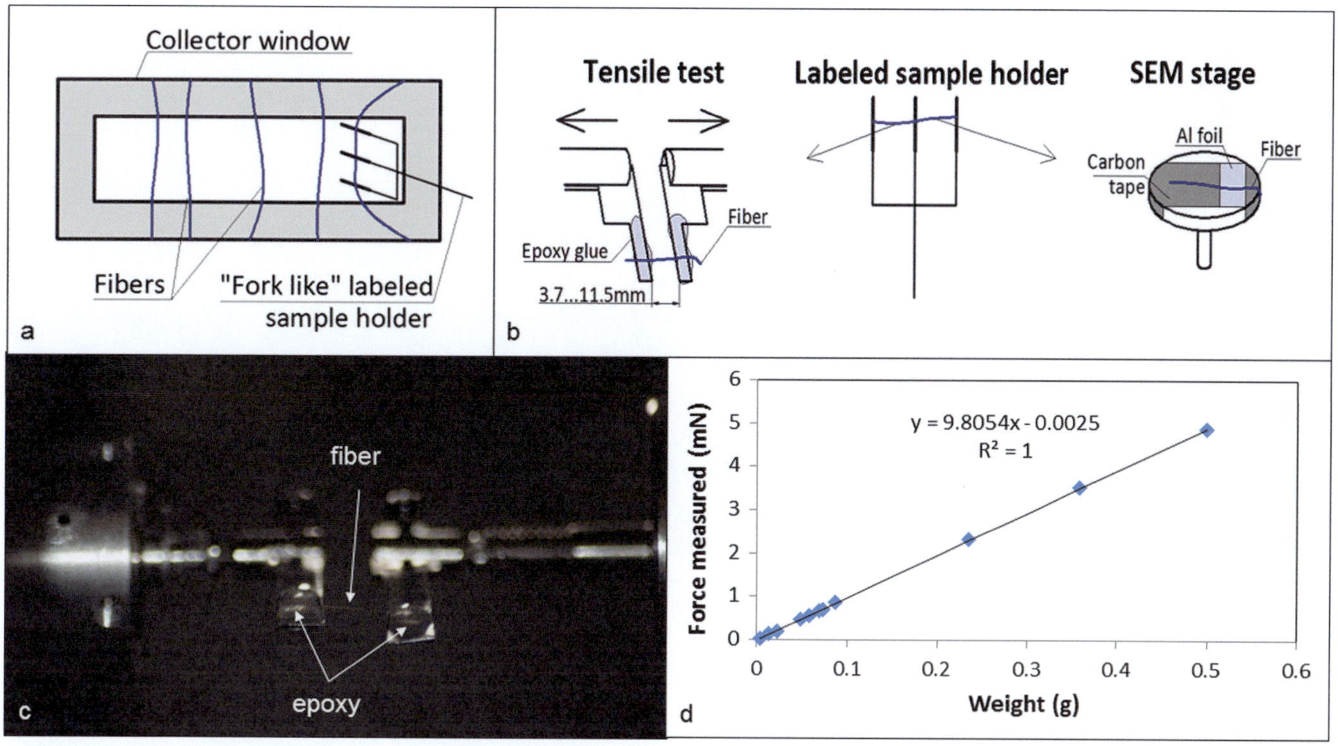

Fig. 6.3 Sample preparation procedure and calibration of the nanomechanics testing system. (**a**) Schematic of the window collector and the procedure for obtaining individual NFs; (**b**) schematic of sample preparation for mechanical testing and diameter measurement; (**c**) NF mounted on the grips of the nanomechanics testing system; (**d**) calibration curve obtained by measuring the force exerted by pre-weighted steel foil samples

was glued using epoxy glue directly to strips of metal preplaced in the testing system grips (see Fig. 6.3b–c). It should be noted that this method prevents on-axis placement of the NFs. Prior to testing of the NFs, the system, in addition to internal calibration, was externally calibrated. Precut pieces of steel foil of different weights were placed on the grip while the system was placed vertically, and the compression force recorded (see results of the calibration test in Fig. 6.3d). The recorded force during the calibration experiment changed by more than two orders of magnitude. The minimum recorded force was ~33 μN, which is approximately an order of magnitude higher than the minimal forces recorded during mechanical tests of NFs.

The relative range of the diameters obtained can be seen in Fig. 6.4a. PAN NFs exhibited pronounced elasto-plastic behavior with large deformations to failure (see typical stress/strain curves in Fig. 6.4b). True stress and true strain were used to describe material behavior at large deformations. Stress/strain curves with lower signal to noise ratios (usually associated with smaller diameters) were smoothed, using a running average smoother with a ten point window. Figure 6.4c shows an SEM micrograph of an edge of a NF broken in tension (not from a specific tensile test), indicating presence of multiple necking in the failure region. Post-mortem examination in an SEM revealed no evidence of slippage of the NF in the epoxy.

Material properties—*i.e.* Young's modulus, maximum stress at failure (strength), strain at failure, and toughness (the area under the engineering stress/strain curve), were extracted and plotted as a function of NF diameter (see Figs. 6.5 and 6.6). Dramatic simultaneous improvements in modulus and strength were observed (see Fig. 6.5), with the largest increases evident for smallest tested NF diameters (<250 nm). The highest observed values were up to an order of magnitude better than those for commercial PAN fibers. While the presence of size effects in these properties is not unusual, and qualitatively similar effects were previously reported in the literature, the magnitude of the reported size effects (approximately two orders of magnitude in both strength and modulus) was significantly higher.

Notably, these dramatic increases came without sacrificing strain to failure, subsequently resulting in maximum toughness approximately an order of magnitude higher than that of best available advanced fibers (Fig. 6.6). This behavior is unusual in engineering materials, which commonly exhibit strength/ductility/toughness trade-off, and highly desirable for safety critical applications. In case of our NFs, however, while engineering strain to failure exhibited large scatter typical of experimental materials, it showed no trend and remained above 50 %.

Structural investigation and comparison of mechanical behavior of as-spun NFs to that of annealed NFs allowed us to propose and verify a structural explanation of this behavior, postulating that the increases in the observed mechanical

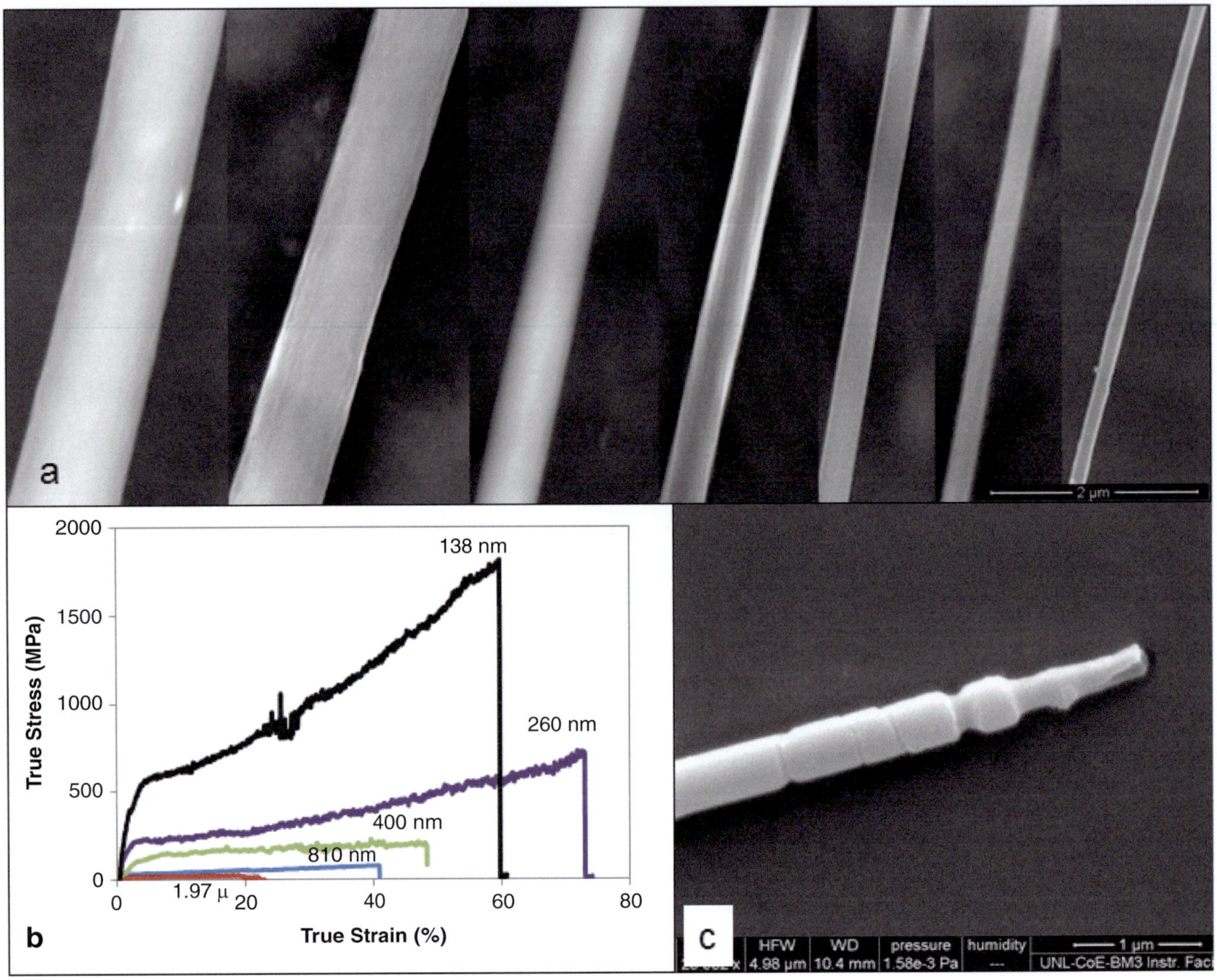

Fig. 6.4 (**a**) Range of diameters obtained; (**b**) typical stress/strain diagrams for different NF diameters; (**c**) an edge of a failed NF showing signs of necking

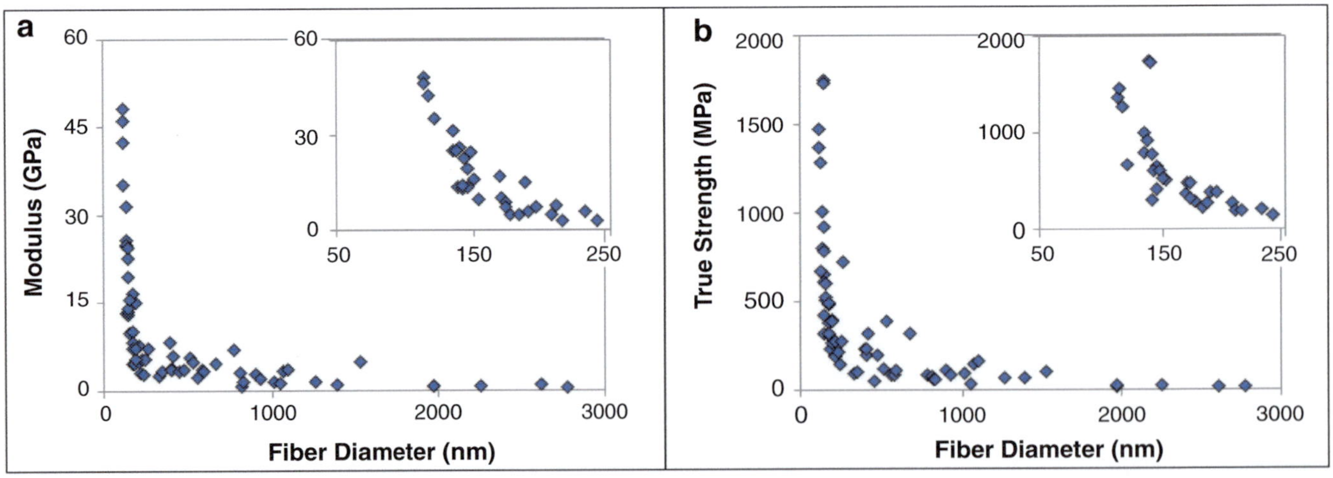

Fig. 6.5 Size effects in (**a**) modulus and (**d**) true strength. *Insets* show magnification of the size effect in the diameter region below 250 nm. Figure modified from [14]

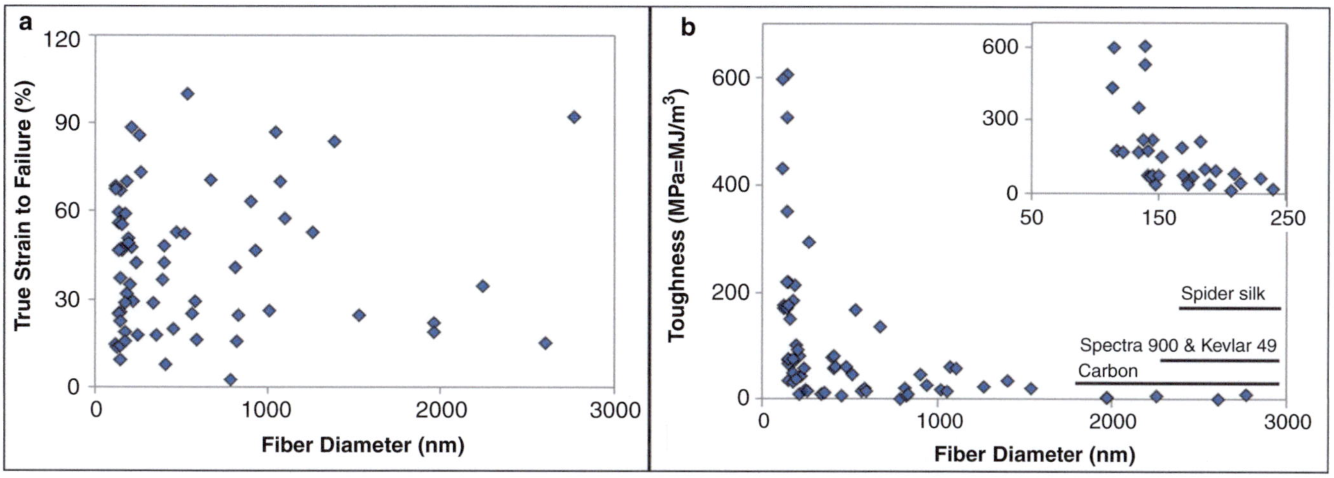

Fig. 6.6 Size effects in (**a**) strain to failure, and (**b**) toughness (lines indicate comparison values for several high performance fibers and spider silk). *Inset* in (**b**) shows magnification of the size effect in the diameter region below 250 nm. Figure modified from [14]

properties are due to improved polymer chain orientation within the amorphous phase of the polymer (for additional details and discussion see [14]).

While previous examinations of other polymer systems focused (either by choice or by necessity) on NF strength and modulus, the described protocol provided a complete picture of mechanical behavior of individual NFs, allowing us to propose a more comprehensive explanation for the observed changes. In addition, these results were obtained for macroscopically long specimens and are readily transferable to macroscopic behavior, while previous examinations, which used short samples require additional considerations for possible length effects.

As a result, the developed protocol of individual NF testing and data reduction provides a benchmark for future studies, allowing for simple comparison between the different studies, and will be instrumental for further optimization of mechanical properties of continuous NFs.

6.5 Future Challenges and Directions

One of the remaining challenges is the mechanical examination of NFs with ultrafine diameters (below those presented above), with the goal of determining both the size effect behavior in that region and ultimate properties achievable. While, nominally, the force resolution of our nanomechanics system would allow for a significant additional reduction of the NF diameter and the measured force, in practice, the signal to noise ratio for individual NFs approaching 100 nm in diameter becomes extremely low, and extraction of information from the stress/strain curves in this diameter range is extremely difficult and introduces large errors.

In order to deal with this issue, a modified protocol was developed in our group and applied to DNA NFs [33]. In this case, when examining NFs with small diameters, multiple fibers of similar diameters were placed in parallel and tested together. The obtained stress/strain curve was then deconvoluted to extract individual stress/strain curves.

The deconvolution process is based on the stiffness analogy to multiple Hookean springs in parallel (see Fig. 6.7a), where the total stiffness of the system at each strain is the sum of stiffnesses of individual NFs. At each failure point the force-extension curve is marked with a vertical jump. The resulting ratio of the force jump value to the force value just before the jump indicates the ratio of the single fiber stiffness versus stiffness of the whole system (see Fig. 6.7b and Eq. (6.1)). Individual NF stiffness can then be extracted by solving the linear equation system created at each jump point. The force-strain curve of the single fibers is reproduced by scaling the force values of the curve by the ratio (see Eq. 6.2).

$$r_i = \frac{\Delta F_i}{F(\varepsilon_i)} \tag{6.1}$$

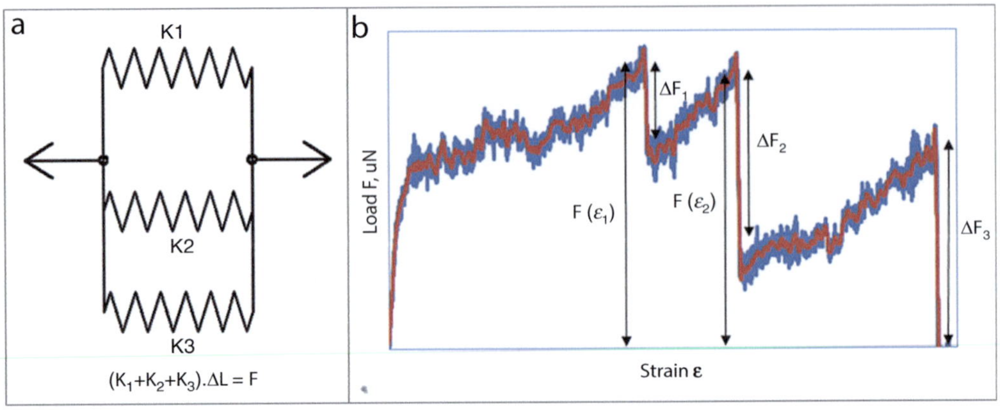

Fig. 6.7 Procedure for deconvolution of simultaneous multiple NF testing. (**a**) Schematic representation of the spring analogy to the multiple fiber stiffness; (**b**) Multiple fiber mechanical test curve (*blue curve* is original data, *red curve* is smoothed) and the parameters needed for curve deconvolution. Figure adopted from [33]

$$\begin{cases} K_1 = (K_1 + K_2 + K_3)\cdot r_1 \\ K_2 = (K_2 + K_3)\cdot r_2 \\ K_3 = \Delta F_3 \end{cases} \rightarrow \begin{cases} F_1(\varepsilon) = F\cdot r_1 \\ F_2(\varepsilon) = F\cdot r_2 \\ F_3(\varepsilon) = F\cdot r_3 \end{cases} \rightarrow \begin{cases} \sigma_1(\varepsilon) = F_1(\varepsilon)/A_1 \\ \sigma_2(\varepsilon) = F_2(\varepsilon)/A_2 \\ \sigma_3(\varepsilon) = F_3(\varepsilon)/A_3 \end{cases} \quad (6.2)$$

This procedure can be extrapolated to a larger number of NFs tested simultaneously and applied to different polymer systems where direct measurements of individual NFs cannot be performed.

6.6 Conclusions

Over the last 10 years, multiple studies performed on individual electrospun NFs have demonstrated strong size effects in properties of these nanofilaments, explaining poor mechanical performance of previously examined NF assemblies. However, many of these studies were not done through failure, providing only a partial picture of NF mechanical behavior, and limiting possible structural explanations. In addition, unknown length effects introduce potential bias in reported results.

The developed mechanical testing and data reduction protocol for macroscopically long NFs described in this paper allows for testing through failure of NFs with large plastic deformations and extraction of a complete picture of their mechanical behavior. It both showed previously unavailable and unique property combinations in electrospun NFs, and is immediately macro relevant because of the size of the specimens examined.

In addition, the described protocol can potentially be coupled with in-situ testing and techniques such as Raman spectroscopy that provide information about the changes in material structure throughout the test. Such a combination can provide currently unavailable information on failure mechanisms, and length effects in mechanical properties. A modified protocol of simultaneous testing of NFs with small diameters allows for examination and extraction of currently unavailable data, extending the examination of size effects in NF mechanical properties into the diameter region, which is currently unexplored due to testing system limitations.

Acknowledgements This work was supported by grants from ONR (N000141410663) and NSF (DMR 1310534).

References

1. Ashby, M. F. Materials selection in mechanical design, 3rd ed.; Butterworth-Heinemann, (2004).
2. The behavior of structures composed of composite materials; Vinson, J.R., Sierakowski, R.L., Eds.; Solid Mechanics And Its Applications; Springer Netherlands: Dordrecht, 2002; Vol. 105.
3. Hearle, J.W.S.: High-performance fibres. Woodhead Publishing Limited, Cambridge (2001)
4. Walsh, P.J. Carbon fibers. In: ASM Handbook, vol. 21 "Composites", (2001)

5. Hiramatsu, T., Nishimura, T.: Recent technological progress of PAN-based carbon fibre. Mater. Des. **10**(2), 93–100 (1989)
6. Dzenis, Y.A.: Structural nanocomposites. Science (80-.). **319**(5862), 419–420 (2008).
7. Ritchie, R.O., Dzenis, Y.A.: The quest for stronger, tougher materials: A letter and a response. Science (80-.). **320**(5875), 448, (2008)
8. Papkov, D.: Size effects in continuous polyacrylonitrile-based polymer, composite, and carbon nanofibers. PhD Dissertation, (2014)
9. Richard-Lacroix, M., Pellerin, C.: Molecular orientation in electrospun fibers: From mats to single fibers. Macromolecules **46**(24), 9473–9493 (2013)
10. Greiner, A., Wendorff, J.H.: Electrospinning: A fascinating method for the preparation of ultrathin fibers. Angew. Chemie Int. Ed. **46**(30), 5670–5703 (2007)
11. Jaeger, R., Schönherr, H., Vancso, G.: Chain packing in electro-spun poly (ethylene Oxide) visualized by atomic force microscopy. Macromolecules **9297**(96), 7634–7636 (1996)
12. Tombler, T., Zhou, C., Alexseyev, L., Kong, J., Dai, H., Liu, L., Jayanthi, C., Tang, M., Wu, S.: Reversible electromechanical characteristics of carbon nanotubes under local-probe manipulation. Nature **405**(6788), 769–772 (2000)
13. Yu, M., Lourie, O., Dyer, M.J., Moloni, K., Kelly, T.F., Ruoff, R.S.: Strength and breaking mechanism of multiwalled carbon nanotubes under tensile load. Science (80-.). **287**(5453), 637–640 (2000)
14. Papkov, D., Zou, Y., Andalib, M.N., Goponenko, A., Cheng, S.Z.D., Dzenis, Y.A.: Simultaneously strong and tough ultrafine continuous nanofibers. ACS Nano **7**(4), 3324–3331 (2013)
15. Pai, C.: Morphology and mechanical properties of electrospun polymeric fibers and their nonwoven fabrics. PhD Dissertation, (2011)
16. Ji, Y., Li, B., Ge, S., Sokolov, J.C., Rafailovich, M.H.: Structure and nanomechanical characterization of electrospun PS/clay nanocomposite fibers. Langmuir **22**(3), 1321–1328 (2006)
17. Wang, M., Jin, H., Kaplan, D.L., Rutledge, G.C.: Mechanical properties of electrospun silk fibers. Macromolecules **37**(18), 6856–6864 (2004)
18. Mack, J.J., Viculis, L.M., Ali, A., Luoh, R., Yang, G., Hahn, H.T., Ko, F.K., Kaner, R.B.: Graphite nanoplatelet reinforcement of electrospun polyacrylonitrile nanofibers. Adv. Mater. **17**(1), 77–80 (2005)
19. Ko, F., Gogotsi, Y., Ali, A., Naguib, N., Ye, H., Yang, G.L., Li, C., Willis, P.: Electrospinning of continuous carbon nanotube-filled nanofiber yarns. Adv. Mater. **15**(14), 1161–1165 (2003)
20. Cuenot, S., Demoustier-Champagne, S., Nysten, B.: Elastic modulus of polypyrrole nanotubes. Phys. Rev. Lett. **85**(8), 1690–1693 (2000)
21. Tan, E.P.S., Lim, C.T.: Physical properties of a single polymeric nanofiber. Appl. Phys. Lett. **84**(9), 1603–1605 (2004)
22. Bellan, L.M., Kameoka, J., Craighead, H.G.: Measurement of the Young's Moduli of individual polyethylene oxide and glass nanofibres. Nanotechnology **16**(8), 1095–1099 (2005)
23. Zussman, E., Chen, X., Ding, W., Calabri, L., Dikin, D.A., Quintana, J.P., Ruoff, R.S.: Mechanical and structural characterization of electrospun PAN-derived carbon nanofibers. Carbon N. Y. **43**(10), 2175–2185 (2005)
24. Arinstein, A., Burman, M., Gendelman, O., Zussman, E.: Effect of supramolecular structure on polymer nanofibre elasticity. Nat. Nano **2**(1), 59–62 (2007)
25. Zussman, E., Burman, M., Yarin, A.L., Khalfin, R., Cohen, Y.: Tensile deformation of electrospun nylon-6,6 nanofibers. J. Polym. Sci. Part B Polym. Phys. **44**(10), 1482–1489 (2006)
26. Jaeger, D., Schischka, J., Bagdahn, J., Jaeger, R.: Tensile testing of individual ultrathin electrospun poly(L-lactic acid) fibers. J. Appl. Polym. Sci. **114**(6), 3774–3779 (2009)
27. Naraghi, M., Chasiotis, I., Kahn, H., Wen, Y., Dzenis, Y.: Novel method for mechanical characterization of polymeric nanofibers. Rev. Sci. Instrum. **78**, 085108(1–7) (2007)
28. Naraghi, M., Arshad, S.N., Chasiotis, I.: Molecular orientation and mechanical property size effects in electrospun polyacrylonitrile nanofibers. Polymer (Guildf). **52**(7), 1612–1618 (2011).
29. Beese, A.M., Papkov, D., Li, S., Dzenis, Y., Espinosa, H.D.: In situ transmission electron microscope tensile testing reveals structure–property relationships in carbon nanofibers. Carbon N. Y. **60**, 246–253 (2013)
30. Griffith, A.A.: The phenomena of rupture and flow in solids. Philos. Trans. R. Soc. London. Series A, Contain. Pap. a Math. or Phys. Character **221**, 163–198 (1921)
31. Weibull, W.: A statistical distribution function of wide applicability. J. Appl. Mech. **18**, 293–297 (1951)
32. Bažant, Z.P., Chen, E.-P.: Scaling of structural failure. Appl. Mech. Rev. **50**(10), 593–627 (1997)
33. Maleckis, K.: Mechanical properties and structure of DNA and collagen nanofilaments. Master's Theses (2012)

Chapter 7
Evaluating Pile-Up and Sink-In During Nanoindentation of Thin Films

MariAnne Sullivan and Bart Prorok

Abstract Determining mechanical properties is crucial when selecting materials for thin film applications. Nanoindentation has been used as an experimental method because of its ease in testing small-scale applications, such as thin films. When the film material properties are complicated by substrate effects, more analysis is necessary to capture the correct information using this technology. In this work, we discuss the differences in pile-up when changing film and substrate. We aim to describe these trends quantitatively through material properties. First, we show our model is an improved method to the nanoindentation technique, and is vital in testing films on substrates. Next, we have measured pile-up and sink-in with a technique including scanning electron microscopy images to better understand the effects of these phenomena on material properties. Also, we have examined sink-in similarly based on projected area measurements with an emphasis on yielding properties. These evaluations lead to further information of film properties that were unknown through nanoindentation in previous tests, which in turn is crucial for thin film applications like electronics and coatings.

Keywords: Nanoindentation • Thin films • Plasticity • Nanoscale mechanics • Pile-up

7.1 Introduction

Nanoindentation is a vital tool in extracting material properties, and has become more important with growing uses of thin films and protective coatings. In order to prevent failure of materials, the properties such as elastic modulus or hardness need to be known before implementing the material in a crucial application such as electronics. Many have studied these types of models, and the one that is most well-known and still used today is the Oliver-Pharr method [1]. This utilizes the basics of the load displacement curve to calculate the stiffness or hardness of materials. Some other have taken it a step further, such as Doerner and Nix [2] and Gao [3], and more recently from Zhou and Prorok [4, 5] to describe thin films on substrates. The components of film and substrate need to be considered together, and are a function of the increasing depth into the surface. Equation 7.1, from Zhou and Prorok, has been successfully used for extracting material properties from film/substrate systems:

$$\frac{1}{E} = \frac{1}{E_f}(1 - \Phi_s)\left(\frac{E_f}{E_s}\right)^{0.1} + \frac{1}{E_s}(\Phi_f) \tag{7.1}$$

Where $\Phi_f = e^{-\alpha_f\left(\frac{t}{h}\right)}$ and $\Phi_s = e^{-\alpha_s\left(\frac{t}{h}\right)}$. The material properties are elastic modulus of the film and substrate, E_f, E_s, and the Poisson's ratios of the film and substrate, α_s and α_f. The variable h is the displacement into surface while t is the film thickness. This equation was utilized in the following work for accurately modeling the composite modulus vs displacement into surface behavior and extracting the film properties. When considering compliant materials on stiff substrates, as was done in this work with gold on ceramic substrates, there arises a problem. Other researchers have tried to correct for pile-up that occurs during nanoindentation, extra area that may not be accounted for when using traditional calculations. One specific method that is modified with our work is done by Kese et al. [6, 7]. While pile-up may affect the extraction of material properties from the Oliver-Pharr equation because it changes the area, the Zhou-Prorok model seemed unaffected. The following work discusses how the pile-up may or may not be affecting the extraction of material properties in these cases where pile-up is present. Sink-in can also be studied in a similar way, and can be used for examining plastic properties.

M. Sullivan (✉) • B. Prorok
Department of Mechanical Engineering, Auburn University, Auburn, AL 36849, USA
e-mail: mzs0050@auburn.edu

B.C. Prorok, L. Starman (eds.), *MEMS and Nanotechnology, Volume 5*,
Conference Proceedings of the Society for Experimental Mechanics Series, DOI 10.1007/978-3-319-22458-9_7

7.2 Experimental Procedure

Thin films on substrates were tested using nanoindentation. First, gold and platinum films were sputtered using a Denton sputtering system with DC power and a rotating sample holder. Conditions were chosen to control film thickness from previous work [4]. A titanium layer of 10 nm was deposited first on all substrates to promote adhesion. This was followed by gold or platinum with the aim of 250, 500, 750, and 1000 nm. The first substrates tested were silicon, because of its well-known properties. However, other ceramics were tested as well, such as aluminum oxide, aluminum nitride, and magnesium oxide. All substrates were chosen as ceramics because they are non-plastically deforming, and would not additionally pile-up, making it solely a film phenomenon. Nanoindentation tests were completed using an MTS Nanoindenter XP system with continuous stiffness measurements, allowing for extraction of properties as a function of depth into the film. An array of 25 indents was used and averaged using the program Analyst. A JEOL 7000 F scanning electron microscopy (SEM) was utilized to image the indents after completion. ImageJ was then used to measure each indent impression. This is discussed further in the results and discussion section.

7.3 Results and Discussion

The methods described were previously used to show that changing film thickness did not affect the extraction of material properties or the indent impression. This is shown in previous work with gold films on silicon substrates [8]. These were very noteworthy results, and to test further, gold films were used on multiple substrates. Film thickness did not play a role in the pile-up, so instead, the substrates were varied to see if the same trends would be visible. First, nanoindentation results from indenting a gold film on 360 nm on all of the ceramic substrates are seen in Fig. 7.1.

These results showed that the model accurately predicted experimental values throughout the entire film thickness for each film and substrate combination. Also, more importantly, when rearranged, the model can extract the film elastic modulus. For all of the samples, this fell around 79GPa, which is expected for the gold films. It also verifies the validity of the model. As the stiffness changes with the substrate, it was thought that there may be a difference in the response of the individual indents. Figure 7.2 shows an indent at the same depth, 400 nm, for all film/substrate combinations. The dashed triangle exhibits the expected triangular impression from the three-sided Berkovich indenter tip. Even though the elastic properties were shown to be different for each sample, the visual impressions of the indents were surprisingly all alike, including the pile-up regions. After comparing visually, a methodology to measure the residual indents was used to apply quantitative information to the test results.

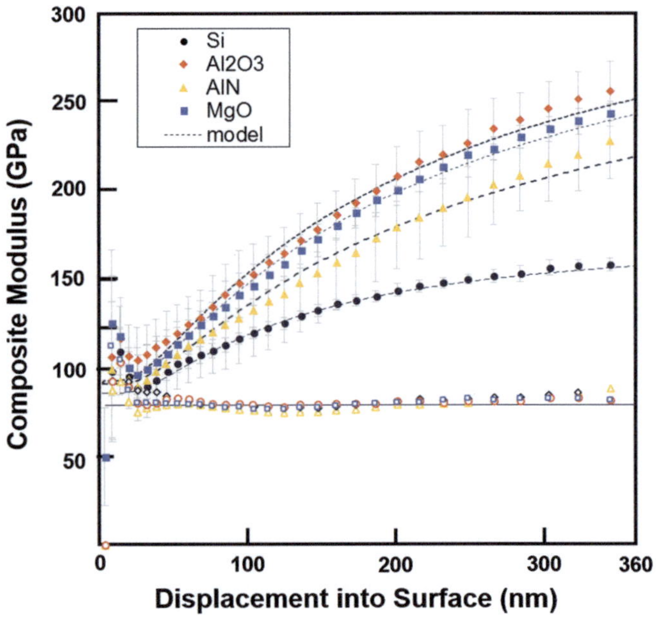

Fig. 7.1 Modulus vs. displacement into surface for gold 360 nm thick on ceramic substrates (*closed symbols*), following the Zhou-Prorok model (*dashed lines*) and extracted film elastic modulus (*open symbols*)

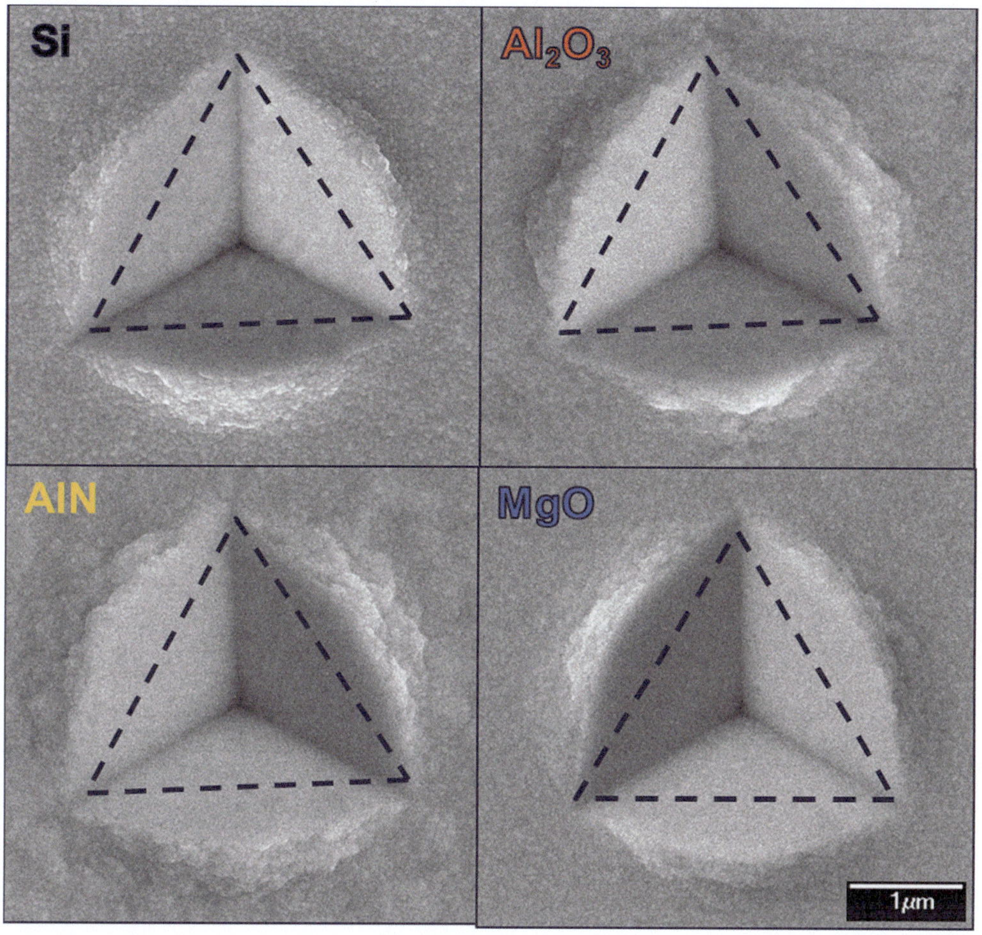

Fig. 7.2 Indent impressions seen after nanoindentation for 360 nm gold on silicon, aluminum oxide, aluminum nitride, and magnesium oxide all at the same indent depth, 400 nm

Previous work described a new method using ImageJ software to measure the residual indent impressions. It is explained further in Fig. 7.3. An indent with pile-up can be measured a few different ways, and the following were chosen. First, the expected Oliver-Pharr triangular area can be measured, and should follow a power law equation because of the self-similar geometry of the indenter tip. Next, the extra area due to pile-up can also be measured, which has been termed pile-up area in this work. Finally, total area would be the addition of these two terms, so both the triangular Berkovich area and also the pile-up area equate to total area. ImageJ software can measure the area based on pixels, and the areas are carefully chosen so as to calculate the correct values. The dashed lines in Fig. 7.3 show the areas that are measured with this method.

Figure 7.4 exhibits these values of measured indent area at increasing indent depths. As the indenter penetrates deeper into the film, all of the measurements increase because more of the material is engaged. The colors correspond between Figs. 7.3 and 7.4, as to which areas are being measured. The pile-up area is smallest, and the total area is largest. The Oliver-Pharr area can be compared to the equation $A = 24.5\,h^2$ because of the geometry of the indenter. The measurements for this area follow the dashed line representing this equation, and further verify the measurements are correct.

With the explained methodology for measured pile-up, the protocol was applied to the images of gold on ceramic substrates from Fig. 7.3. The projected areas were identical, and for clarify solely the pile-up area is depicted in Fig. 7.5. The plot indicates the pile-up area as a function of indent depth. All the films were the same thickness, 360 nm, so this is shown with a vertical dashed line. There was an average of four indents measured per indent depth, and shown is the average with standard deviation represented with error bars. Even with varying stiffness responses, shown from the nanoindentation data, the pile-up is identical across the samples at a specific indent depth. This further shows that the pile-up is independent of the substrate material, and instead, is solely dependent on indent depth into the film. As only gold films were measured, the next step was to see how changing the film would affect the type of pile-up response.

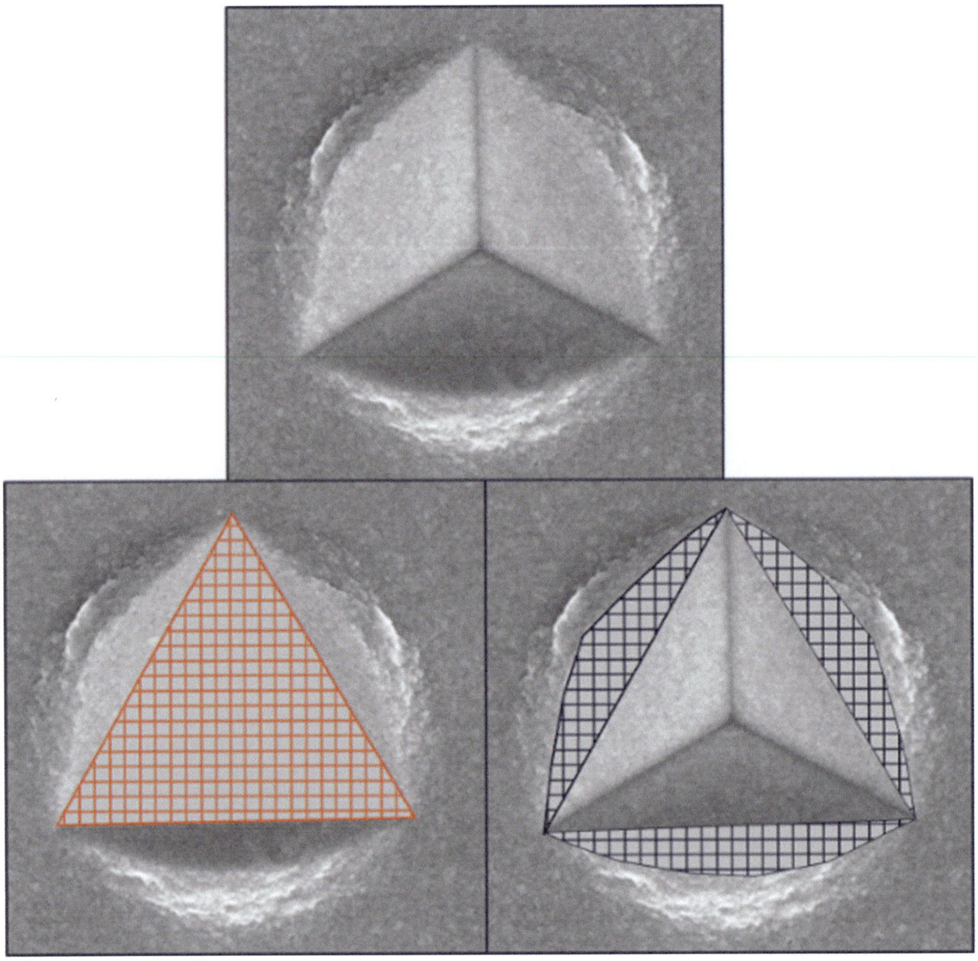

Fig. 7.3 And indent with pile-up and Image J projected areas for Oliver-Pharr area (*red*) and pile-up area (*dark blue*)

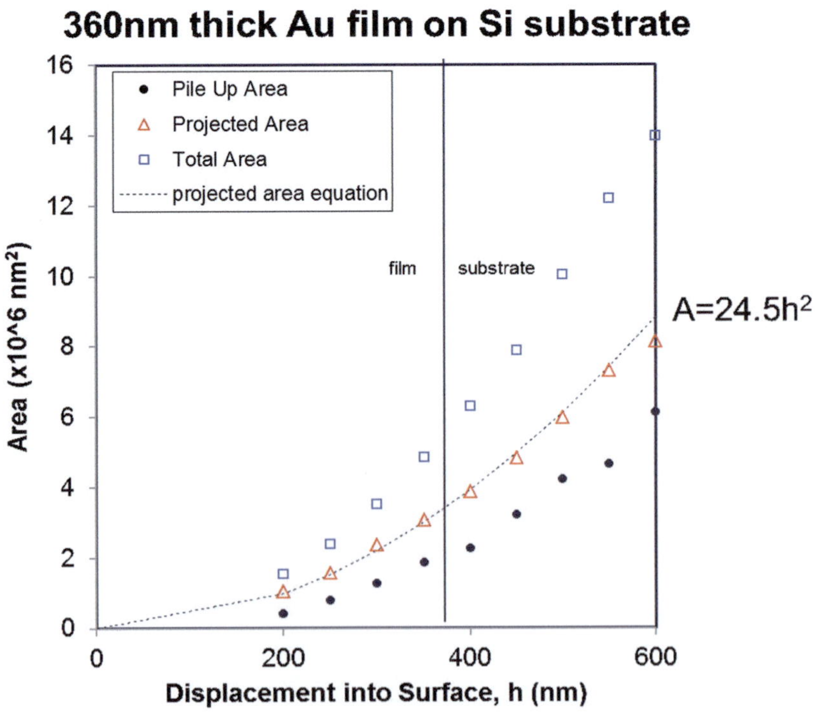

Fig. 7.4 Area measurements for pile-up, projected Berkovich area, and total area, all following similar increasing trends

Fig. 7.5 Pile-up area as a function of displacement for all gold films on ceramic substrates, following the same pattern no matter the film/ substrate combination

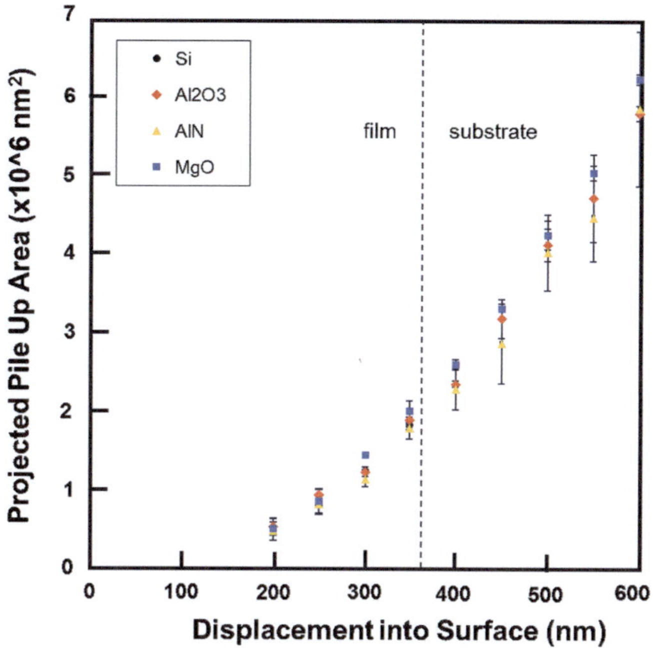

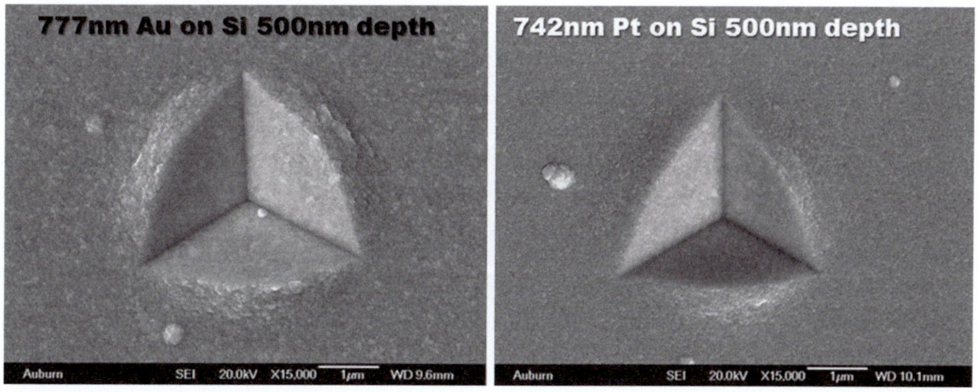

Fig. 7.6 Pile-up difference for gold and platinum films

The next set of experiments then tested platinum on silicon as a different film material with a very different elastic modulus of 168GPa as compared to the 79GPa for gold. Films were deposited with similar thicknesses to the previously tested gold films, and indented in the same manner as described in the experimental procedure. The images seen in Fig. 7.6 show that pile-up can be different for different films, even with the same substrate materials. This exhibits differences in plasticity of these materials that may be affecting the growth of the pile-up during nanoindentation. In Fig. 7.6, indents at the same indent depth, 400 nm, are imaged on two film/substrate combinations, and the pile-up appears different. This is unlike the situation where film was kept the same, but substrate was changed. The ImageJ measurement system was then used to numerically compare the pile-up of gold or platinum films, which is seen in Fig. 7.7. Open markers indicate platinum of different film thicknesses on the same substrate. Closed markers represent gold with different thicknesses on the silicon substrate. For both gold and platinum, no matter the film thickness, the pile-up responds identically and in a growing manner. However, there is a difference in the amount of pile-up, as platinum has consistently less at each indent depth. This can be attributed to plastic properties of the film that are varying. Future work will be to find the reason for the amount of pile-up and how it is affecting the residual indent. The goal is to discover what is physically happening at this scale.

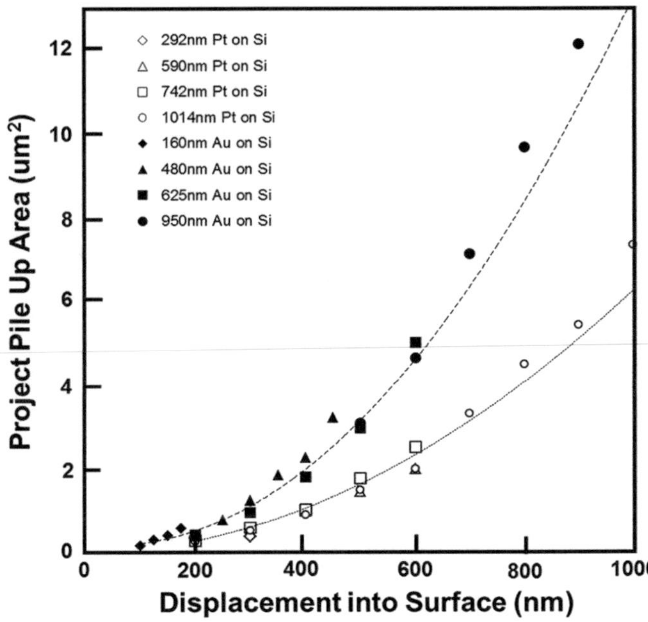

Fig. 7.7 Measured pile-up vs displacements into surface for gold (*closed markers*) and platinum films (*open markers*)

7.4 Conclusions

Overall, a few noteworthy findings are presented in this work. First, the Zhou-Prorok model could correctly measure elastic properties despite changes in pile-up. Pile-up was found to be independent of the substrate elastic properties, as proven with gold on four different non-plastically deforming ceramic substrates. This was completed with a novel technique for measuring pile-up area. Because pile-up did not change with substrate material, the film was then varied. The pile-up was different for gold and platinum films, but still followed individual increasing trends. Understanding how the material properties relate to the amount of pile-up will help future researchers to account for it, or know when it is not necessary to account, such as the case with the elastic properties found in this work. Future work will dissect the plasticity portion of pile-up to hopefully extract further properties such as yield stress from nanoindentation tests.

References

1. Oliver, W.C., Pharr, G.M.: An improved technique for determining hardness and elastic modulus using load and displacement sensing indentation experiments. J. Mater. Res. **7**, 1564–1583 (1992)
2. Doerner, M.F., Nix, W.D.: A method for interpreting the data from depth-sensing indentation instruments. J. Mater Res. **1**, 601–609 (1986)
3. Gao, H., Chiu, C.-H., Lee, J.: Elastic contact versus indentation modeling of multi-layered materials. Int. J. Solids Struct. **29**, 2471 (1992)
4. Zhou, B., Prorok, B.: A discontinuous elastic interface transfer model of thin film nanoindentation. Exp. Mech. **50**, 793–801 (2010)
5. Zhou, B., Prorok, B.C.: A new paradigm in thin film nanoindentation. J. Mater Res. **25**, 1671–1678 (2010)
6. Kese, K., Li, Z.C.: Semi-ellipse method for accounting for the pile-up contact area during nanoindentation with the Berkovich indenter. Scr. Mater. **55**, 699–702 (2006)
7. Kese, K.O., Li, Z.C., Bergman, B.: Method to account for true contact area in soda-lime glass during nanoindentation with the Berkovich tip. Mater. Sci. Eng. A **404**, 1–8 (2005)
8. Sullivan M., Prorok B.C.: Newly discovered pile up effects during nanoindentation. In: The Society for Experimental Mechanics, Greenville, SC, 2014

Chapter 8
Tapered Cantilevered Bimorphs for Piezoelectric Energy Harvesting: Characterization with Impedance Spectroscopy

Naved A. Siddiqui, Dong-Joo Kim, Ruel A. Overfelt, and Barton C. Prorok

Abstract Cantilevered piezoelectric bimorphs resonating in the fundamental mode have previously been explored for energy harvesting, where it has been reported that tapering the geometry into triangular shapes provides a constant strain distribution over the surface of cantilevered bimorphs, unlike conventional rectangular geometries with a large tip mass. This enhances the k_{31} electromechanical coupling coefficients in the devices, an aspect which is studied using impedance spectroscopy in this paper; a tool that can measure the exact values for the optimal load resistance which can be applied to maximize power output from piezoelectric bimorphs. It is found that tapering the devices into triangular geometries lowers the internal impedance of the device at the resonance frequency; and increases the impedance of the device at the anti-resonance frequency when compared to rectangular counterparts with the same device volume. The peak value for the phase angle shift is also larger in triangular bimorphs, indicating a larger dielectric loss at the median of the resonance and anti-resonance frequencies; implying that the devices here would be more compliant, i.e. provide a wider resonance frequency bandwidth. Therefore, the advantages provided by altering geometry to improve the strain distribution in the devices are reflected here from an electronic standpoint.

Keywords: Energy Harvesting • Piezoelectric • Cantilevered bimorph • Impedance spectroscopy • Electromechanical coupling

8.1 Introduction

With the quest of supplying a localized source of energy for low powered sensor nodes and various other devices where batteries are difficult to integrate, piezoelectric energy harvesting is an area that has been the subject of extensive research over the past decade [1–7]. This phenomenon typically utilizes cantilevered piezoelectric bimorphs undergoing base excitations, designed to resonate in the fundamental mode, thereby creating axial strains in the device due to the transverse displacements. Therefore, the d_{31} mode of operation is exploited while generating a continuous charge. Typically, cantilevered piezoelectric bimorphs are rectangular in shape with a proof mass attached at the free end of the beam, which causes a stress concentration near the fixed end of the beam. It has been shown in studies that tapering the surface of the cantilevered bimorph results in a constant strain profile along the surface of the beam due to the constantly changing area moment of inertia [8]. This provides lower maximum stresses, and more reliable operation of the bimorph, along with enhanced electromechanical coupling coefficients [9–11].

It has also been well documented that the peak power produced by a piezoelectric bimorph into a load resistor connected in series has optimal value(s) [11–14]. For low coupled harvesters, there exists one value for the optimal load resistance, while in high coupled harvesters, there exist two such values [14, 15]. These peaks occur when the internal impedance of the device is matched with the external load resistance [16]. In other words, with impedance (Z) (defined as $|Z| = R + jX$), being a complex (j) quantity, impedance matching occurs when the load resistance (R) is purely resistive, and there is no reactance (X). At such values, the magnitude of the phase angle is zero.

This paper discusses the effects of varying geometry and proof masses on cantilevered piezoelectric bimorphs by making use of impedance spectroscopy. Using this method, the exact values at which impedance matching can occur can be

---

N.A. Siddiqui (✉) • D.-J. Kim • R.A. Overfelt • B.C. Prorok
275 Wilmore Laboratories, Materials Engineering, Auburn University, Auburn, AL 36849, USA
e-mail: naved.siddiqui@auburn.edu

B.C. Prorok, L. Starman (eds.), *MEMS and Nanotechnology, Volume 5*,
Conference Proceedings of the Society for Experimental Mechanics Series, DOI 10.1007/978-3-319-22458-9_8

determined by utilizing simple frequency response measurements, which play an important role in characterizing the piezoelectric energy harvesters. By making use of these measurements, the electromechanical coupling coefficient is determined using the equation $k^2 = 1 - (f_r/f_a)^2$ [17], where f_r is the resonance frequency, and f_a is the anti-resonance frequency. The resonance frequency is the frequency where the impedance is minimum, and the anti-resonance frequency is the frequency where the impedance is maximum.

8.2 Methodology

The piezoelectric bimorphs used in this study are the PZT-5H materials obtained from Piezo Inc. These are of the T220-H4-series that have 0.19 mm thick piezoelectric layers, and a 0.13 mm thick brass layer. A rectangular bimorph is prepared with an overhang length of 21.5 mm and thickness of 6.35 mm, which results in a short-circuit resonance frequency of 515 Hz for a 70 mm^3 device. A corresponding isosceles triangular device that has a matching resonance frequency and device volume is calculated to be 30.5 mm in altitude, 8.95 mm in base width (and 0.51 mm total thickness). Interestingly, the resonance frequency of the triangular device matches that of its rectangular counterpart when the length-to-width ratios match. Details on this modal analysis are present in previous papers from this author [18–20].

An impedance analyzer, Agilent 4294A is utilized to measure various dielectric properties as a function of frequency, with a constant oscillating voltage. A 100 mV oscillating voltage is sufficient for high frequency measurements, but in low frequency applications, the source voltage can be increased up to a volt to improve signal-to-noise ratios. Prior to measurements, the instrument must be calibrated under both open circuit, and short circuit conditions. A frequency sweep is chosen for the dielectric response such that a resolution of ~0.5 Hz is obtained (for example, 800 data points for a frequency sweep from 350 Hz to 750 Hz). Various responses as a function of frequency can be obtained such as Z-θ, where Z is the impedance measured in ohms, and θ is the phase angle. Another response that can be measured is C_p-D, where C_p is the parallel capacitance of the piezoelectric device, and D is the dissipation factor, which is an inverse of the loss tangent. All these values provide peaks at specific frequencies, which primarily relate to the resonance and anti-resonance frequency of the device.

8.3 Results and Discussion

The two devices as calculated and prepared in the previous section are tested with changing values of proof mass. In the first case, no proof mass is attached, and in the next two cases, 2 g blocks of stainless steel are added. The impedance spectroscopy data, where the values of impedance and phase angle are observed as a function of frequency are consolidated in Table 8.1 for impedance values at the resonance (f_r) and anti-resonance frequency (f_a). The peak phase angle, at the median of the resonance and anti-resonance frequency is also presented. The corresponding k_{31} electromechanical coupling coefficient is also calculated. This table is thoroughly discussed in this paper.

Table 8.1 Impedance spectroscopy data for the 70 mm^3 rectangular and triangular piezoelectric bimorphs with varying proof masses

Sample	f_r (Hz)	|Z| (Ohms)	f_a (Hz)	|Z| (Ohms)	Peak Phase°	k_{31}
Rect—0 g	511.2	5558	530.5	143,641	51	0.272
Tri—0 g	501	3023	530	223,137	66	0.326
Rect—2 g	120	37,172	130.2	246,245	18.3	0.388
Tri—2 g	95.3	17,165	102	1,192,406	75.5	0.356
Rect—4 g	86.8	24,392	90.9	1,650,976	65.8	0.297
Tri—4 g	67.8	27,326	72.4	3,373,502	69.0	0.372

8.3.1 No Proof Mass

The devices, as calculated and mentioned in the methodology section are illustrated in Fig. 8.1. The impedance spectra for these two devices, without a proof mass are shown in Fig. 8.2a and b respectively. It can be seen that the resonance frequency of both devices on the dielectric frequency response plots occur at 511 Hz and 501 Hz (slight error from calculated values for triangular devices), which is where the impedance of the two devices are at their minimum levels. The corresponding phase angle at this frequency is zero, meaning that the impedance here is purely resistive. Therefore, maximum power can be extracted when the external load resistor matches this impedance, which for the rectangular bimorph is found to be 5.56 kΩ, and for the triangular device is found to be at a lower value of 3.02 kΩ. In terms of the second peak, i.e. the maximum impedance values, which occur at the anti-resonance frequencies, it can be see that the rectangular device has a maximum internal impedance of 143.6 kΩ and the triangular device this time has a higher internal impedance of 223.1 kΩ. Therefore, the triangular devices provide a bigger difference in magnitude between their minimum and maximum impedance values, which relates to a larger calculated electromechanical coupling coefficient, as calculated and shown in Table 8.1. The k_{31}

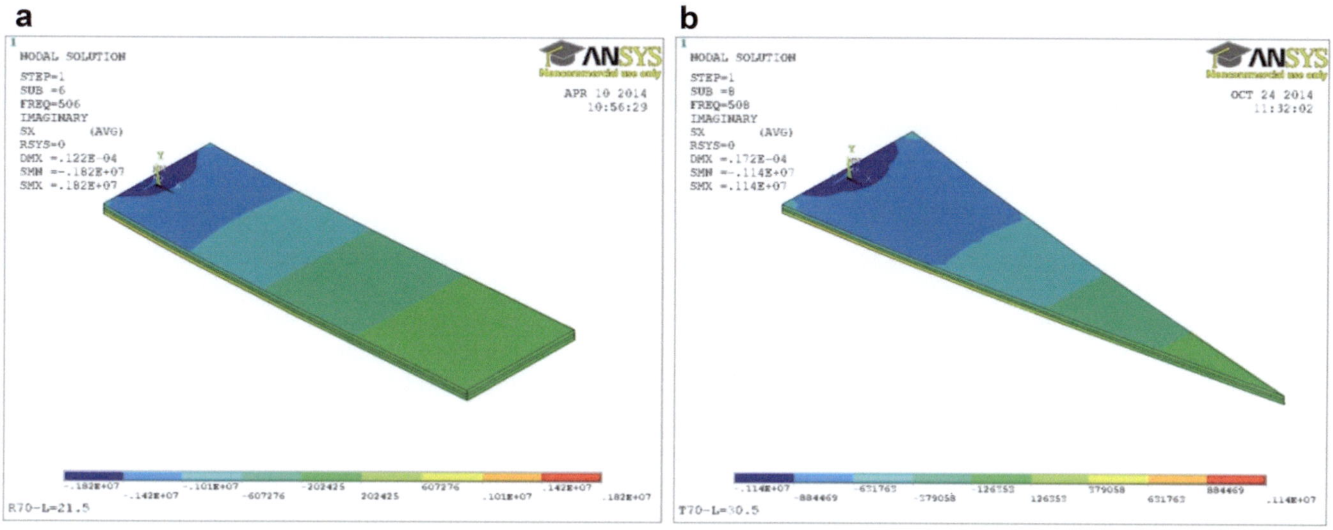

Fig. 8.1 (**a**) Rectangular bimorph having dimensions of l = 21.5 mm, w = 6.35 mm and t = 0.51 mm; v = 70 mm^3. (**b**) Rectangular bimorph having dimensions of l = 30.5 mm, w = 8.95 mm and t = 0.51 mm; v = 70 mm^3

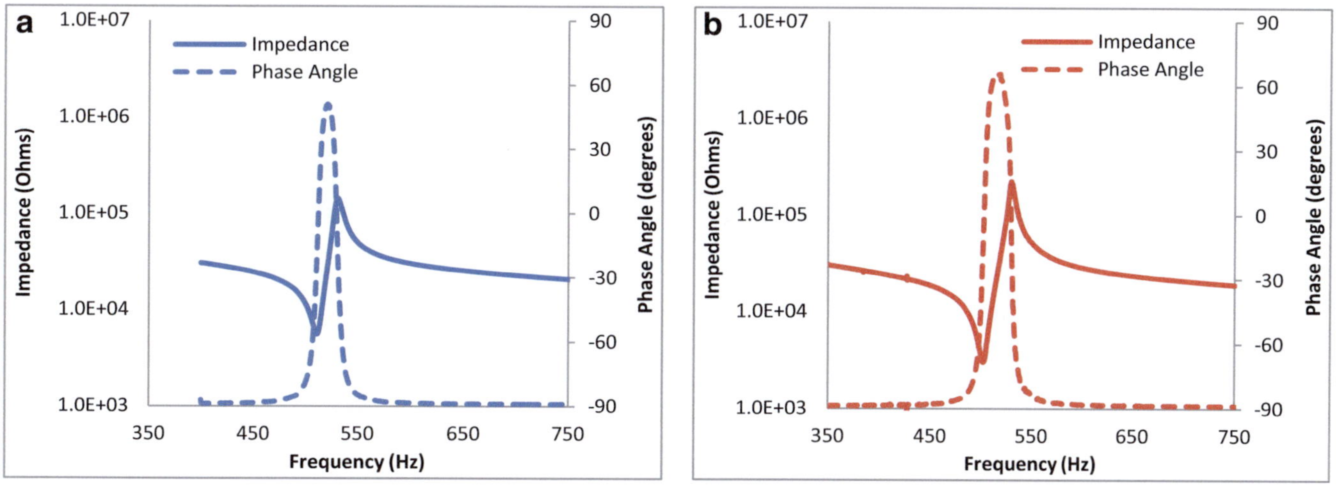

Fig. 8.2 (**a**) Impedance-Phase Angle for 70 mm^3 rectangular bimorph. (**b**) Impedance-Phase Angle for 70 mm^3 triangular bimorph

value for the rectangular bimorph is found to be 0.27, and for the triangular device, with the same volume, similar resonance frequency, and inertial loading is found to be 0.33.

Another important aspect is the peak magnitude of the phase angle. This peak is directly related to the dielectric loss in the system at the given frequency. It can be clearly seen that the triangle without the proof mass gives a much larger peak phase angle of 66° compared to 51° for the rectangle. Therefore, while the electromechanical coupling in the triangular device is larger, the dielectric loss at the frequency that is the median of the resonance and anti-resonance frequency is larger. While this larger dielectric loss diminishes the power produced at these frequencies, the larger resulting damping ratios can provide a wider frequency bandwidth for power generation.

8.3.2 2 g proof mass

The addition of a nominal 2 g proof mass, made of up 316 Stainless Steel key-stocks, having dimensions of 6.35 × 6.35 × 6.35 mm^3 is placed on the bimorphs, as depicted in Fig. 8.3. The addition of the 2 g has an order of magnitude effect on the amount of power generated [19, 20], and also lowers the resonance frequency of the devices to much more usable levels, of around 100 Hz in the case of the 70 mm^3 devices as designed. However, in doing so, the comparability between the two devices in terms of matching short circuit resonance frequencies is lost (see Table 8.1). The mismatch in the frequencies is also apparent in Fig. 8.4, where the impedance peak shift can be seen.

The magnitudes of these peaks, as presented in Table 8.1 are important to discuss. It can be noticed that the impedance at the resonance frequency, i.e. the minimum impedance for the rectangular and triangular devices go up by an order of magnitude when compared to the previous case where there was no proof mass. The resonance–impedance for the rectangular device with the 2 g proof mass is 37.2 kΩ, while the resonance–impedance for the triangular device is 17.2 kΩ. The anti-resonance–impedance, i.e. the impedance maxima's are also substantially higher where the rectangular device has a maximum impedance of 246.2 kΩ and the triangular device has a maximum impedance of 1192.4 kΩ. Therefore, the trend is similar here, where the triangular device has a much larger difference between the minimum and maximum values. The magnitude of the peak phase angle for the triangular device is also large compared to the rectangular counterpart, indicating a larger dielectric loss (thus a large damping ratio/dissipation factor). However, the calculated k_{31} value for the triangular bimorph is shown to be lower here at 0.36, compared to the rectangular bimorphs 0.39. Due to the fact that the calculation of the two devices is based on the resonance and anti-resonance frequency, where the two resonance frequencies do not match; this comparison may not be entirely viable.

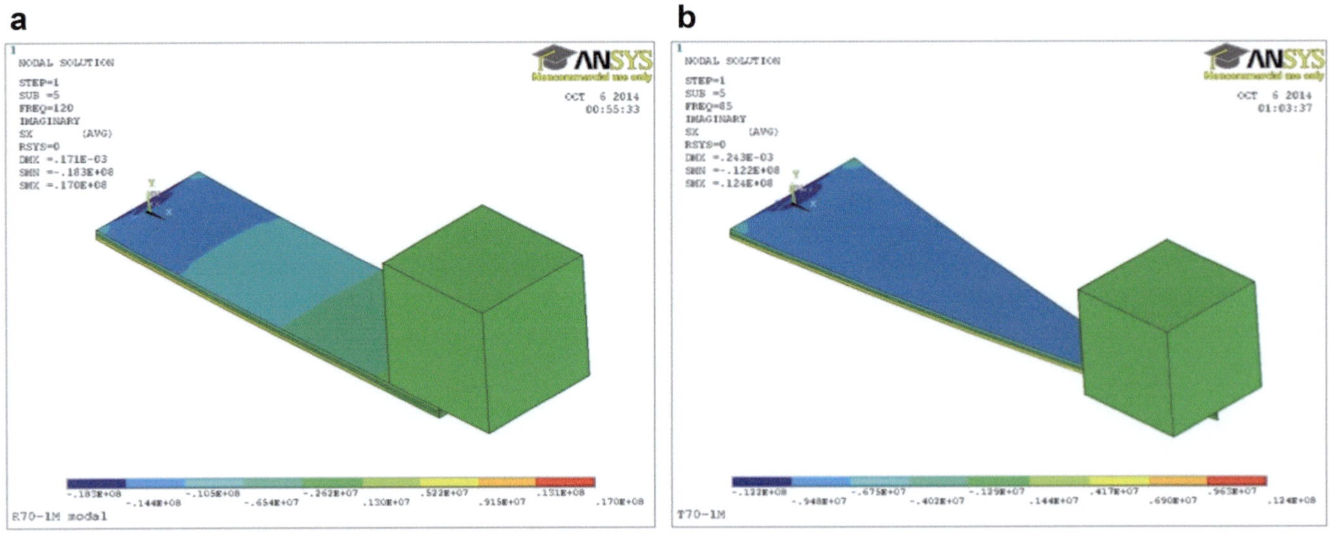

Fig. 8.3 (**a**) 70 mm^3 rectangular bimorph with a 2 g proof mass. (**b**) 70 mm^3 triangular device with a 2 g proof mass

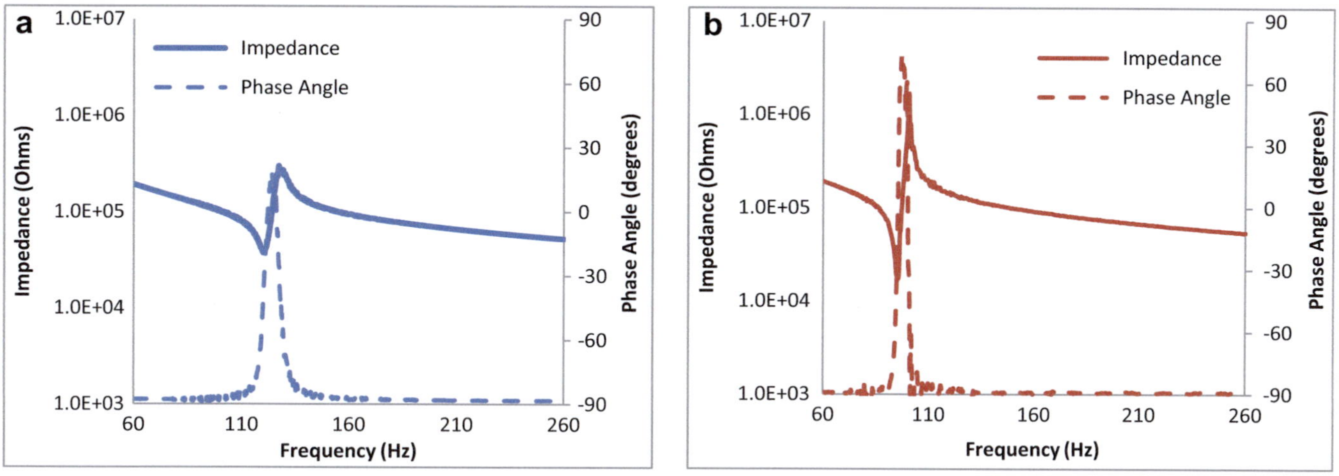

Fig. 8.4 (**a**) Impedance-Phase Angle for 70 mm^3 rectangular bimorph with 2 g proof mass. (**b**) Impedance-Phase Angle for 70 mm^3 triangular bimorph with 2 g proof mass

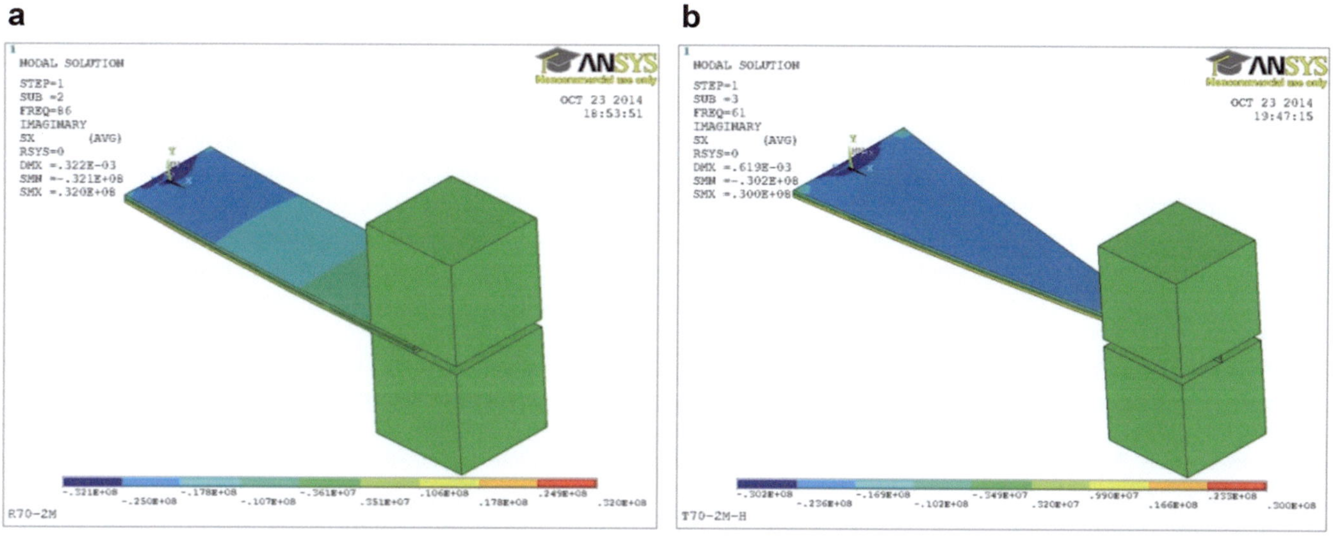

Fig. 8.5 (**a**) 70 mm^3 rectangular bimorph with a 4 g proof mass. (**b**) 70 mm^3 triangular bimorph with a 4 g proof mass

8.3.3 4 g Proof Mass

A further 2 g of proof masses, as illustrated in Fig. 8.5 were added to see the effect on power output [18–20], and also to see if the resonance frequencies of the two devices would converge. As it can be seen in Fig. 8.6, this was not the case, as the resonance frequency of the rectangular device 87 Hz, and the triangular device is 68 Hz, with 4 g of proof mass. It was also seen that the optimal load resistance values were converging to a single point rather than two distinct peaks. Here, impedance spectroscopy data shows that the impedance minima are still within the same range as in the last case with the 2 g proof mass (24.4 and 27.3 kΩ for the rectangular and triangular devices respectively), but the impedance maxima (at the anti-resonance frequency) are shifting to very large values of 1.7 and 3.4 MΩ respectively. Generating power with such large impedances would be very expensive due to large amounts of Joule heating. However, the improved electromechanical coupling coefficient, and larger difference between the impedance maxima and minima from the triangular device is again seen.

It is also worth mentioning that the impedance spectroscopy data for the devices with the 4 g proof masses have resonance frequencies that are close to the limit of the Agilent 4294A impedance analyzer, which has a lowest possible frequency of 40 Hz. In order improve the signal-to-noise ratio, the oscillating voltage here was increased to 1 V, since the data with low oscillating voltages produced a large amount of noise (Figs. 8.5 and 8.6).

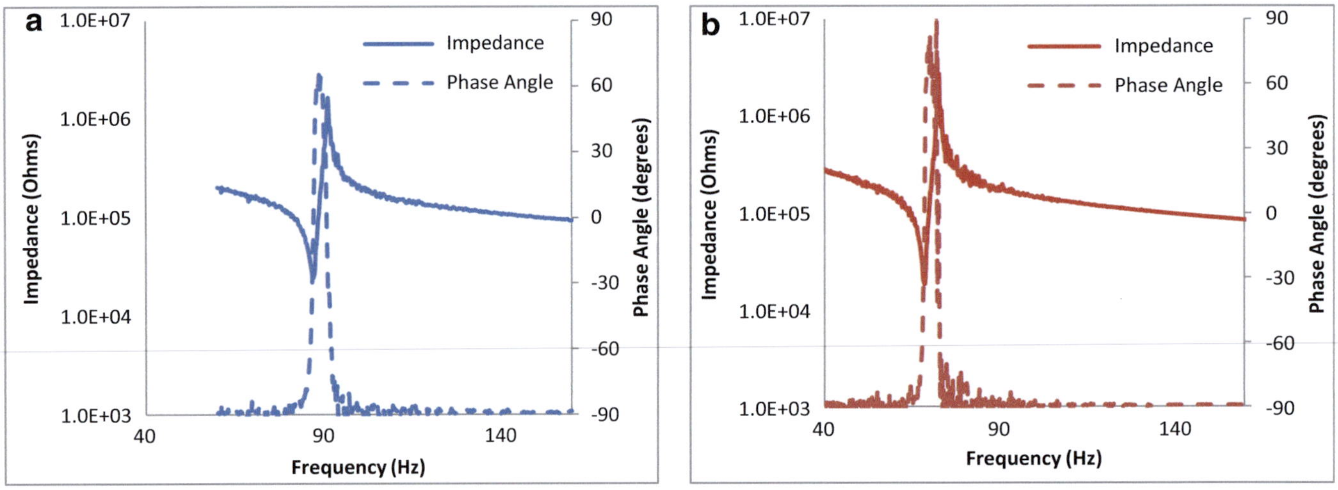

Fig. 8.6 (**a**) Impedance-phase angle for 70 mm^3 rectangular bimorph with 4 g proof mass. (**b**) Impedance-Phase Angle for 70 mm^3 triangular bimorph with 4 g proof mass

8.4 Conclusions

Impedance spectroscopy can be a very useful tool to characterize the electronic characteristics of piezoelectric bimorphs, wherein the exact values for the internal impedance of the devices can be measured. Such data would be very meaningful for circuitry applications into which the vibrating bimorph would be attached. In previous studies, it has been shown that tapering the geometry into triangular bimorphs provides a constant strain profile, and lowers the maximum stress at the fixed end of the cantilevered bimorphs, thereby providing an enhanced electromechanical coupling coefficient, an aspect that is confirmed in this study. In this study, with enhanced electromechanical coupling coefficients, it is seen that the impedance at the resonance frequency for triangular bimorphs is lower than rectangles, and the impedance at the anti-resonance frequency is substantially larger. The lower impedance at the resonance frequency is a parameter of interest, and therefore tapering the geometry to improve mechanical strain distributions is reflected from an electronic standpoint as well.

Acknowledgements This work was partially supported by the U.S. Federal Aviation Administration (FAA), Office of Aerospace Medicine under Cooperative Agreements 04-C-ACE and 07-C-RITE.

References

1. Sodano, H.A., Inman, D.J., Park, G.: A review of power harvesting from vibration using piezoelectric materials. Shock Vibration Digest **36**, 197–205 (2004)
2. Beeby, S.P., Tudor, M.J., White, N.M.: Energy harvesting vibration sources for microsystems applications. Measure. Sci. Technol. **17**, R175–R195 (2006)
3. Priya, S.: Advances in energy harvesting using low profile piezoelectric transducers. J. Electroceram. **19**, 167–184 (2007)
4. Anton, S.R., Sodano, H.A.: A review of power harvesting using piezoelectric materials (2003–2006). Smart Mater. Struct. **16**, R1–R21 (2007)
5. Cook-Chennault, K.A., Thambi, N., Sastry, A.M.: Powering MEMS portable devices—a review of non-regenerative and regenerative power supply systems with special emphasis on piezoelectric energy harvesting systems. Smart Mater. Struct. **17**, 043001 (2008)
6. Khaligh, A., Zeng, P., Zheng, C.: Kinetic energy harvesting using piezoelectric and electromagnetic technologies—state of the art. IEEE Trans. Indust. Electron. **57**, 850–860 (2010)
7. Li, H., Tian, C., Deng, Z.D.: Energy harvesting from low frequency applications using piezoelectric materials. Appl. Phys. Rev. **1**, 041301 (2014)
8. Roundy, S., Leland, E.S., Baker, J., Carleton, E., Reilly, E., Lai, E., Otis, B., Rabaey, J.M., Wright, P.K., Sundararajan, V.: Improving power output for vibration-based energy scavengers. IEEE Pervasive Comput. **4**, 28–36 (2005)
9. Goldschmidtboeing, F., Woias, P.: Characterization of different beam shapes for piezoelectric energy harvesting. J. Micromech. Microeng. **18**, 104013 (2008)
10. Dietl, J.M., Garcia, E.: Beam shape optimization for power harvesting. J. Intell. Mater. Syst. Struct. **21**, 633–646 (2010)

11. Xiong, X., Oyadiji, S.O.: (2013) Modal electromechanical optimization of cantilevered piezoelectric vibration energy harvesters by geometric variation. J. Intell. Mater. Syst. Struct. p. 1045389X13502872, (2013)
12. Roundy, S., Wright, P.K., Rabaey, J.: A study of low level vibrations as a power source for wireless sensor nodes. Comput. Commun. **26**, 1131–1144 (2003)
13. Roundy, S., Wright, P.K.: A piezoelectric vibration based generator for wireless electronics. Smart Mater. Struct. **13**, 1131–1142 (2004)
14. Goldschmidtboeing, F., Wischke, M., Eichhorn, C., Woias, P.: Parameter identification for resonant piezoelectric energy harvesters in the low- and high-coupling regimes. J. Micromech. Microeng. **21**, 045006 (2011)
15. Lei, A., Xu, R., Borregaard, L.M., Guizzetti, M., Hansen, O., Thomsen, E.V.: Impedance based characterization of a high-coupled screen printed PZT thick film unimorph energy harvester. J. Microelectromech. Syst. **23**, 842–854 (2014)
16. Kong, N., Ha, D.S., Erturk, A., Inman, D.J.: Resistive impedance matching circuit for piezoelectric energy harvesting. J. Intell. Mater. Syst. Struct. **21**, 1293–1302 (2010)
17. Isarakorn, D., Briand, D., Janphuang, P., Sambri, A., Gariglio, S., Triscone, J., Guy, F., Reiner, J., Ahn, C., De Rooij, N.: The realization and performance of vibration energy harvesting MEMS devices based on an epitaxial piezoelectric thin film. Smart Mater. Struct. **20**, 025015 (2011)
18. Siddiqui, N.A., Roberts, M.I., Kim, D.-J., Overfelt, R.A., Prorok B.C.: Tapered piezoelectric devices for vibration energy harvesting. In: SPIE Smart Structures and Materials + Nondestructive Evaluation and Health Monitoring, 2014, pp. 90572B–90572B-7.
19. Siddiqui, N.A., Kim, D.-J., Overfelt, R.A., Prorok, B.C.: Shape optimization of cantilevered devices for piezoelectric energy harvesting. In: MEMS and nanotechnology, Volume 8 : Springer International Publishing, 2015, pp. 17–24.
20. Siddiqui, N.: Understanding effects of tapering cantilevered piezoelectric bimorphs for energy harvesting from vibrations. PhD, Materials Engineering, Auburn University, Auburn, (2014)

Chapter 9
Time and Temperature Dependence of Stress Relaxation in Al and Al Alloy Thin Films Application for MEMS

A.-W. Huang, C.-H. Lu, C.-C. Wu, and M.-T. Lin

Abstract Metal thin films are using as capacitance switches in microelectromechanical systems (MEMS). The better the thin films can resist relaxation, the longer the lifetime of the MEMS device. Al thin films have been used as capacitance switches, but they have only a weak resistance to relaxation and therefore a shorter lifetime. Solid strengthening can improve the mechanism of the material without decreasing other mechanisms. We added Mg to Al thin films in order to increase the stress relaxation resistance. The viscoelastic behavior of pure aluminum thin film in comparison with thin films of different percentages of Mg is investigated by using bulge testing. Adding Mg in pure Al thin films significantly decrease the relaxation behavior and increase the mechanism of thin film. The result shows the more Mg content in Al thin films the more resistance of thin films. The results of experiments show the normalize modulus decreasing less with a greater content of Mg. This result proves that adding more different atoms can obstruct the movement of dislocation and enhance the mechanism of Al thin films. It shows that Al-Mg thin films have better relaxation resistance than Al thin films and thus serve as a better material for MEMS.

Keyword: Optical methods • Thin films • Mechanical properties of thin films • Optical measurements • Interferometer

9.1 Introduction

The micro-systems technology has grown rapidly during the last decade; with advances in the technology behind micro-systems the need to enhance performance and cost efficiency has increased demand for further miniaturization. Thin films have been used in a variety of applications as structural components, so the mechanical properties of the thin films are very important to the potential reliability of the devices. Thin films final microstructure can effect and be used to explain the mechanical properties. Metal thin films have either a columnar grain structure or a finer microstructure with smaller equiaxed grains. There are two parameters that can control the final microstructure of the thin film on rigid substrate: one is the deposition process and the other is the subsequent thermal treatment.

Thin metal films are critical elements in a many nano- and micro- fabricated technologies. Accordingly, the mechanical properties of micron and nano-scale thin films have become an item of particular interest.

In bulk material such as steel and aluminum has little relaxation and small strain. On the contrary, the fine small grain material such as thin film has be strong influence on relaxation and creep. Many research report that several thin film material is time dependent deformation as the result of relaxation and creep. Yan et al [1] reported that a 27 % reduction in effective elastic modulus occurs under constant strain conditions at room temperature over a 3 day period. RF MEMS switch operating temperature range is surpassed room temperature. Hence, the high temperature has an influence on the relaxation behavior [2, 3]. McLean et al. [4] report that the relaxation is strong temperature dependent in the temperature range of 20–80 °C. Therefore, increasing temperature means increasing both the magnitude and rate of stress relaxation and this situation make RF switch easy to failure at high temperature.

Metal thin films are often using as capacitance switches in MEMS device. Al thin film has low resist to stress relaxation and creep. The goal of this experiment is to add foreign atoms in pure Al thin films to improve the mechanism of Al thin films. These foreign atoms can hinder the movement of dislocation and decrease recrystallize effect on the thin films. Mg is the material that adding in thin films as foreign atoms. The Al-Mg alloy thin films are testing and hope to increase the resist of stress relaxation. If the mechanisms of Al-Mg alloy thin films being increase and the other mechanisms are not decreases, this alloy thin films can be replaced to pure Al thin films to increase the reliability and expand the useful time of MEMS.

A.-W. Huang • C.-H. Lu • C.-C. Wu • M.-T. Lin (✉)
Graduate Institute of Precision Engineering, National Chung Hsing University, 250, Kuo-Kuang Road, Taichung 40227, Taiwan, People's Republic of China
e-mail: mingtlin@nchu.edu.tw

B.C. Prorok, L. Starman (eds.), *MEMS and Nanotechnology, Volume 5*,
Conference Proceedings of the Society for Experimental Mechanics Series, DOI 10.1007/978-3-319-22458-9_9

9.2 Experimental

9.2.1 *Sample Design and Fabrication*

The single test chip is designed (Fig. 9.1) to fit in an apparatus where one side of the paddle plate is pulled by an electrostatic field. The measured strain was converted through the capacitance measurement for the deflection of the cantilever. The samples used in gas pressure bulge testing require a special design. Figure 9.2 shows the standard semiconductor fabrication methods used to make the bulge. The thin film membrane is made with a 20 mm by 20 mm silicon die which has a window 12 mm long and either 2 or 3 mm wide in the center as shows in Fig. 9.2. Each silicon die is coated with 210 nm SiN on both sides. The SiN on the back side contains a 12 mm by 2 mm or 3 mm window made with standard photolithography methods. The silicon dies are immersed in 45 %(weight percent) KOH at a temperature of 85 °C until the silicon is fully etched. Then Al and Al alloy thin films are deposited by DC magnetron sputtering on the fully etched silicon dies.

The sample fabrication procedure is using standard clean room processing. The sample is fabricated on a standard 4 in. diameter double polished silicon wafer. The detail of the fabrication process sequence is shown in Fig. 9.3. The structure can be metallized from both sides. No processing is required after the metal has been deposited. Later, a metallic conducting surface is required on both sides of paddle for stable deflection and capacitance measurements to be made. All the thin films are created by DC magnetron sputtering using a system obtained from AJA international (ATC 2200 sputtering system). The sputter has three guns which can carry three elemental targets. In this experiment, we only used two guns to create the samples. In this experiment, we used weight measurement to determine the deposition rate of Al and Al alloy thin film.

9.2.2 *Testing System*

We used bulge testing to determine time and temperature dependence of viscoelastic stress relaxation in Al and Al alloy thin films. Bulge systems as shows in the figure have two chambers. One is the vacuum chamber; the other is the bulge chamber.

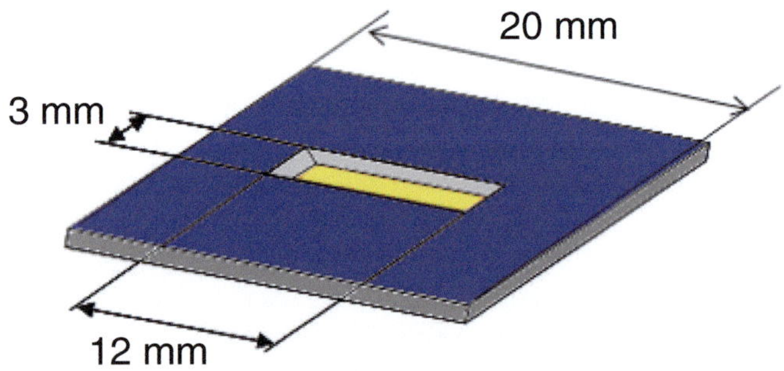

Fig. 9.1 Schematic view of the sample

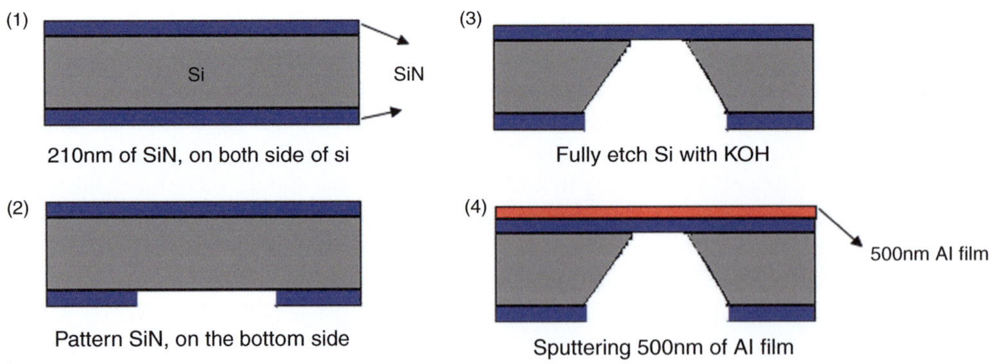

Fig. 9.2 The detail of the fabrication process sequence

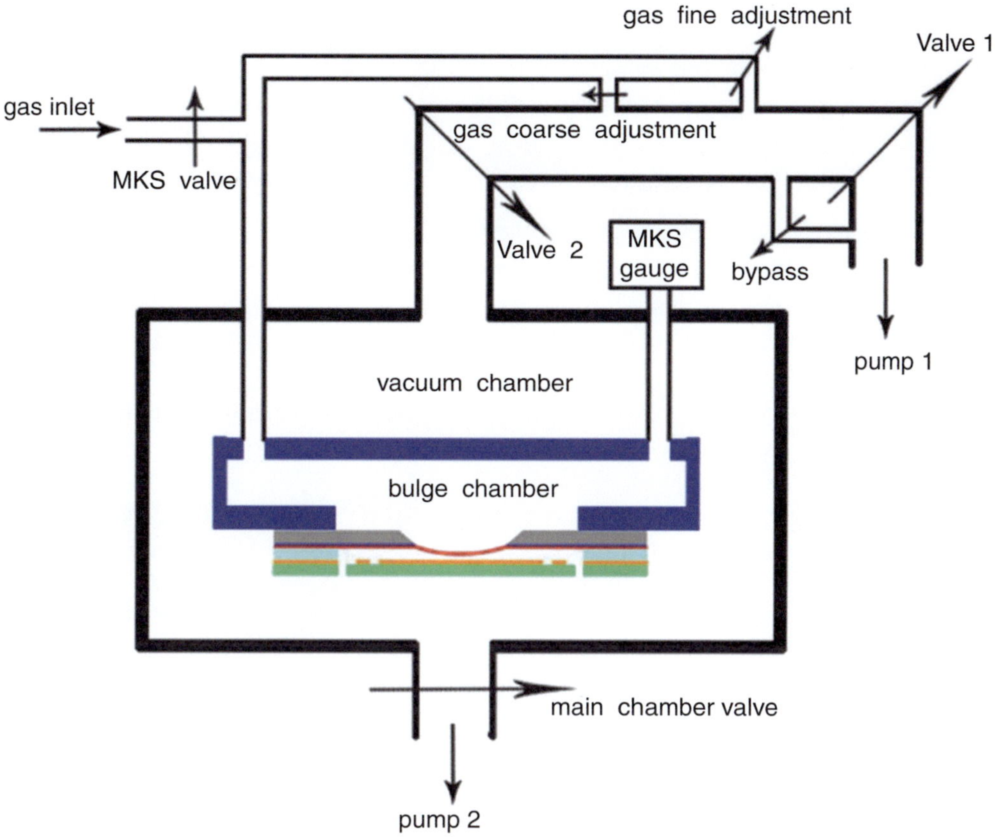

Fig. 9.3 Schematic of the measuring system

There are eight heaters embedded inside the bulge testing system. Those heaters are using to elevate the temperature. In this experiment, the temperature is 50 °C, 70 °C and 80 °C. After creating a vacuum in the chamber, the pressure of N_2 gas is applied to the bulge chamber. Therefore, inside the chamber, one side of the sample membrane is in a vacuum while the other side of sample membrane is under pressure which is created by the application of N_2 gas as shown in Fig. 9.3 [5] and Fig. 9.4 [6].

The applied pressure is controlled by a MSK pressure controller under the command of a computer running the Labview program. The MSK controller obtains the desire pressure as measured by the MSK pressure gauge. The gas pressure creates the pressure difference between both sides of the film and able to bulge the thin membrane. The height of the bulge provides information about the strain in the thin film. Once bulge height is obtained, the radius R can be calculated using the following equation

$$H = R - \sqrt{R^2 - \frac{D^2}{4}}. \tag{9.1}$$

$10The stress in the films then can be calculated as the hoop stress acting on a thin-walled cylinder given by $\sigma = PR/t$ where t is the film thickness. The strain can also be computed from

$$\varepsilon = \frac{2R \sin^{-1}\frac{D}{2R} - D}{2a}. \tag{9.2}$$

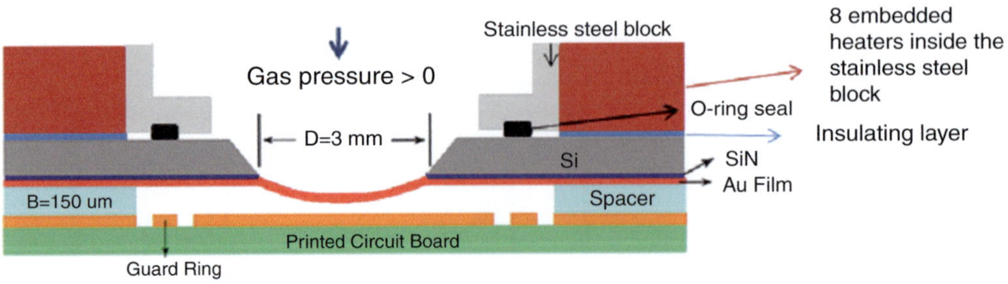

Fig. 9.4 Schematic of loading on a sample, it shows the measurement of load if stress relaxation of the sample occurs during the experiment

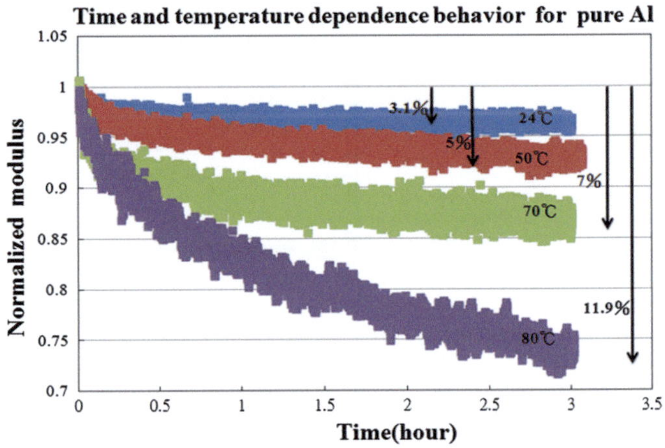

Fig. 9.5 Three hours modulus decay of pure Al thin films at each temperature

9.3 Results

The stress relaxation in thin film membranes has a huge effect on the lifetime of capacitance switches application for MEMS. Hence, the goal of this experiment was to improving the lifetime of capacitance switch application for MEMS by using a promising material with less stress relaxation over the temperature range. In this study, there are three types of thin films tested here, we used: pure Al thin film, 10 % Mg-Al alloy thin film, and 15 % Mg-Al thin film. We tested these thin films to study their stress relaxation effect by using bulge testing over a temperature range of 20–80 °C. Temperature has great influence on the relaxation behavior of the films. During the stress relaxation, the activation energy of the mechanism played the key role. When the temperature is different, the stress relaxation behavior is also different. In this experiment, we tested the Al/SiN composite thin films. When the relaxation behavior of thin film tested reached to its the steady state condition at 24 °C, we raise the temperature to 50 °C, and then to 70 °C, finally ends the test at 80 °C. Each temperature required many runs until it reaches the steady state. Figure 9.5 shows the normalize modulus of Al/SiN composite thin film at the steady state for each temperature. This result shows great temperature effect on the relaxation behavior of the thin films. The percentage of the 3 h normalizes modulus decrease of Al thin film is putting together to be compared. When the temperature increases, the normalize modulus decreases from 3.1 % to 11.9 %. Figure 9.6 shows the normalize modulus of the relaxation for 10 % Mg Al thin film. The 3 h modulus decay of Al alloy thin films at 24, 50, 70, and 80 °C are 2.8 %,4 %,5 %and 6.5 % respectively.

Figure 9.7 shows the result of normalize modulus for various temperatures after fitting the Prony series model. The 3 h modulus decay of the Al alloy thin films at 20 °C, 50 °C, 70 °C, and 80 °C are 2.4 %, 3.3 %, 5 %, and 5.5 % respectively.

In all plots, it shows that when temperature increases, the normalize modulus decreases more. The values of 3 h normalize modulus decays are showing in Table 9.1. The plots show that adding increasing amounts of Mg to Al thin films results in an apparently smaller relaxation decay. While the 3 h modulus decay of pure Al thin films over the 20–80 °C temperature range vary from 3.1 to 11.9 %, for 10 % Mg content they decay 2.8–6.5 % and for 15 % Mg content they decay 2.4–5.5 %. This experiment proves that alloying thin films can enhance the thin films mechanism and increase the endurance of relaxation. The plots also show when Al thin films contain more Mg, the stress relaxation behavior of the film reduced.

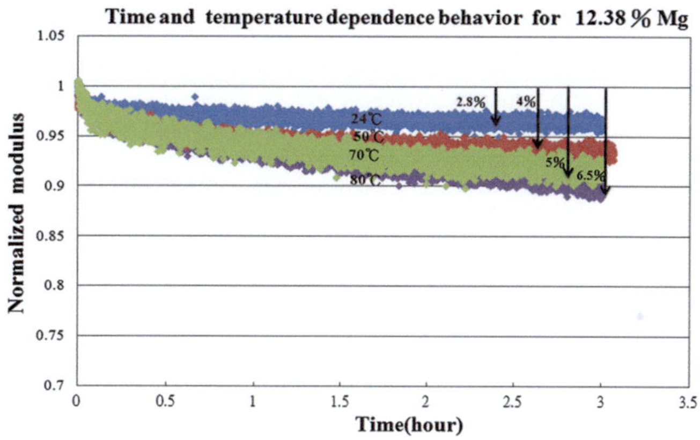

Fig. 9.6 Three hours modulus decay of 10 %thin films at each temperature

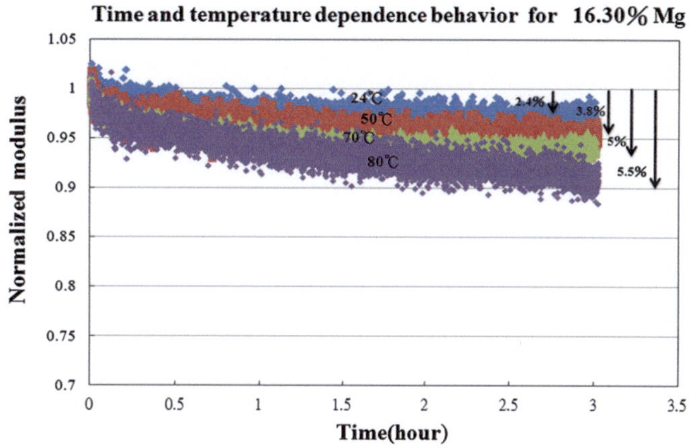

Fig. 9.7 Three hours modulus decay of 15 % thin films at each temperature

Table 9.1 Three hours modulus decay of pure Al, 12.63 % Mg, 16.30 % Mg at different temperature

	3 h decay (%)		
Temperature	Al	12.63 % Mg	16.30 % Mg
24 °C	3.1	2.8	2.4
50 °C	5	4	3.8
70 °C	7	5	5
80 °C	11.9	6.5	5.5

9.4 Conclusions

The viscoelastic behavior of pure aluminum thin film in comparison with thin films of 10 %Mg and 15 % Mg are investigated by using bulge testing. Adding Mg in pure Al thin films significantly decrease the relaxation behavior and increase the mechanism of thin film. The result shows the more Mg content in Al thin films the more resistance of thin films. The mechanism of thin films has been improve to using in RF switch MEMS. However, the electric properties of thin films are also important. In the future work, the Mg-Al alloy thin films have to do the electric test. If the electric properties of Mg-Al thin films are still good, this thin films can use to replace pure Al thin films as RF switch in MEMS device.

References

1. Lakers, R.: Viscoelastic materials. Cambridge University Press, New York (2009)
2. Yan, X., Brown, W.L., Li, Y., Papapolymerou, J., Palego, C., Hwang, J.C.M., Vinci, R.P.: J. Microelectromech. Syst. **18**(3), 570–576 (2009)
3. Kalkman, A.J., Verbruggen, A.H., Janssen, G.C.A.M.: Rev. Sci. Instrum. **74**(3), 1383–1385 (2003)
4. Hyun, S., Brown, W.L., Vinci, R.P.: Appl. Phys. Lett. **83**(21), 4411–4413 (2003)

5. McLean, M., Brown, W.L., Vinci, R.P.: J. Microelectromech. Syst. **19**(6), 1299–1308 (2010)
6. Vlassak, J.J., Nix, W.D.: Mater. Res. **7**(12), 3242–3249 (1992)
7. Kittisun Mongkolsuttirat, time and temperature dependence of viscoelastic stess relaxation in Au and Au alloy thin films, paper.
8. Weihs, T.P., Hong, S., Bravman, J.C., et al.: Mechanical deflection of cantilever microbeams: A new technique for testing the mechanical properties of thin films. J. Mater. Res. **3**(5), 931–42 (1988)
9. Nix, W.D.: Metallurgy. Trans. **20A**, 2217 (1989)
10. Schweitz, J.A.: Mechanical characterization of thin films by micromechanical Techniques. MRS Bull. **17**(7), 34–45 (1992)

Chapter 10
Detecting Interconnect Damage in Shock Using Acoustic Emission Detection

W. Carter Ralph, Elizabeth Elias Benedetto, Gregory N. Morscher, Keith Newman, and Julie Silk

Abstract Acoustic emission detection was used to detect damage events on electronic assemblies during shock testing. The boards were instrumented with an accelerometer and four acoustic transducers and dropped multiple times at acceleration levels between 100 and 250 g. High intensity acoustic events were identified and the origins of the events were located. Dye stain failure analysis was used to identify the type and location of damage between the interconnection and the circuit board, and showed good agreement with the acoustic events. The results indicate that acoustic emission detection can be applied to this test method with the potential to significantly improve its precision and throughput time.

Keywords: Acoustic emission • Dynamic testing • Shock • Damage detection • Electronics

10.1 Introduction

Detection of the initiation of damage in the solder interconnections between components and the printed wiring board has been a challenge for the electronics industry. The current state of the art for both shock and bend conditions combines strain gage measurements, electrical continuity monitoring, and destructive post-test failure analysis. These processes are imprecise and time-consuming, and are infeasible on fully-populated production parts. Acoustic emission detection (AED) has been shown to be a practical tool for improved detection of crack initiation in transient bend tests. This study has extended this technique to the high strain rate mechanical conditions of shock and indicates that AED has the potential to improve the precision, speed, and versatility of crack detection in this test.

Acoustic techniques have been used for decades to detect sub-surface flaws in parts and structures and to detect fracture events during loading. AED is a passive acoustic technique that uses transducers to measure sound waves on a surface and calculate the origin of the event from the position of the sensors, the time delay between the arrival of the events at the sensors, and the velocity of the waves through the medium. The first published study of the use of AED for the electronics industry was in 2011 in which two acoustic sensors were used to detect fracture events in four point bend [1]. This allowed events to be located in one dimension and matched with the time, strain, and displacement measurements. Subsequent studies used four sensors to detect fractures in four point bend [2] and in the industry-standard spherical bend test [3–5] and to locate the events in two dimensions across the board surface. These later studies correlated the relative acoustic intensity of the events to the type of event in order to differentiate between the typical pad crater failure mode at the electrical interconnection and the non-critical glass composite damage. These studies demonstrated how the precise moment of interconnect damage initiation can be identified; that the damage events can be related to strain, load, and displacement conditions on the board; and that the precision and speed of the transient bend test method is greatly improved through the use of AED.

W.C. Ralph (✉)
Southern Research Institute, 757 Tom Martin Drive, Birmingham, AL 35211, USA
e-mail: ralph@southernresearch.org

E.E. Benedetto
Hewlett-Packard, Houston, TX, USA

G.N. Morscher
The University of Akron, Akron, OH, USA

K. Newman
Hewlett-Packard, Palo Alto, CA, USA

J. Silk
Keysight Technologies, San Francisco, CA, USA

B.C. Prorok, L. Starman (eds.), *MEMS and Nanotechnology, Volume 5*,
Conference Proceedings of the Society for Experimental Mechanics Series, DOI 10.1007/978-3-319-22458-9_10

This led to the question of whether AED could be applied to shock testing. There are several challenges to using AED in shock. Transient bend testing is relatively slow (about 0.5 s to first failure) compared to shock testing (2 ms input pulse), which means that the acoustic events are much closer together in time, potentially making these events impossible to differentiate from each other. Due to its dynamic nature, shock induces much more noise from the fixturing and the instrumentation wires, which could mask the acoustic events of the interconnection damage. In addition, maintaining adequate connection between the acoustic sensors and the board surface during the test, without causing additional acoustic interference from the transducers and their wires, is a concern.

This paper presents a study in which the acoustic methods developed for transient bend were modified and applied to shock tests up to 200 g. Multiple drops were performed on a number of test boards over a range of acceleration levels. Acoustic events were detected in all tests and their locations were plotted relative to the location of the electronic component. These events were compared to interconnect damage locations found by post-test failure analysis. Interconnect damage events were differentiated from other events, and the acoustic characteristics were used to predict failure locations that were verified by failure analysis.

10.2 Test and Data Reduction Methods

The test specimens were PBGA packages reflowed onto a test PCA as shown in Fig. 10.1. The packages were 35 $\times$ 35 mm and incorporated organic substrates with fully-encapsulated die. The packages were mounted to the printed circuit assemblies using metal defined pads with 1 mm pitch SAC305 solder balls and SAC305 solder paste. The circuit boards were 203 $\times$ 203 mm, 1.6 mm thick, 8 layers, FR-4 laminate materials. The laminates were fabricated with alternating 0/90 glass cloth orientations. Daisy-chain networks and strain gage lead soldering points were designed into the PCAs and specimens at each corner.

IPC/JEDEC-9703, Mechanical Shock Test Guidelines for Solder Joint Reliability (March 2009), provides a detailed procedure for development of mechanical drop and shock testing conditions. Critically, the mechanical shock test conditions should relate back to anticipated product use conditions, and be correlated through controlled validation studies. Typically, selected shock test conditions are chosen among the eight standardized service conditions specified in JEDEC JESD22-B110B, Mechanical Shock—Component and Subassembly (July 2013). These service conditions are defined by acceleration pulse shape (half-sine), peak acceleration level and pulse duration, and range from 100 g, 2.0 ms to 2900 g, 0.3 ms. For this early-development study, however, neither a specific end-product configuration nor a use environment were defined. Rather

Fig. 10.1 Test board

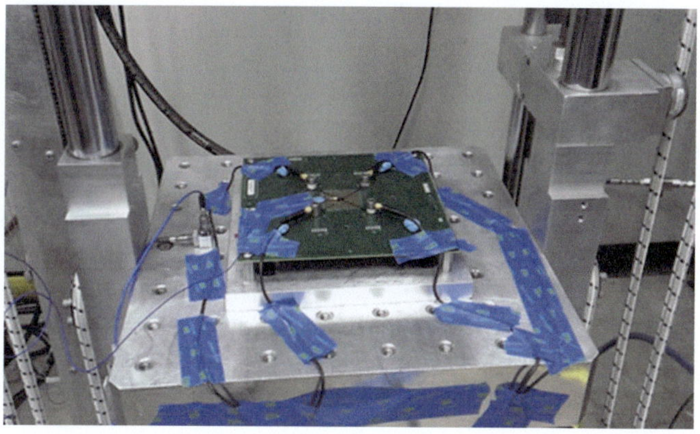

Fig. 10.2 Shock test setup

Table 10.1 Preliminary shock test results

Shock level (g)	Number of drops	Dye stain results
100	1	No opens
100	5	No opens
150	1	No opens
250	1	2 partial cracks
350	1	Partial and full cracks across component
450	1	Full opens across most of component

than select a mechanical shock test condition that correlated to a use environment, the shock levels were determined based primarily upon their ability to generate laminate pad craters in the test assemblies. The selected shock levels roughly matched JEDEC JESD22-B110B service conditions C (100 g, 2.0 ms), D (200 g, 1.5 ms) and E (340 g, 1.2 ms). The pulse duration and half-sine pulse shape were controlled by damping pads, calibrated at each acceleration level. The shock setup is shown in Fig. 10.2. The boards were mounted with the package facing down so that the interconnections are loaded in tension first. A multi-axis accelerometer was mounted on the shock table to measure the input conditions, and another small single axis accelerometer was adhered to the center of the test board to measure the output condition. The acoustic transducers were mounted at the corners, 38 mm from the center of the package, using a thin layer of wax. In order to keep the instrumentation wires from slapping the board during impact, wires were taped to the board with masking tape and the point of contact with the board was secured with putty.

In order to determine the shock level at which pad craters would occur, several boards were put through the drop shock test prior to acoustic emission testing. These boards were evaluated using layout dye stain testing and the results were used to focus subsequent acoustic testing shock levels. The results from these preliminary tests are shown in Table 10.1. From these results, it was determined that 100 g was a "safe" level at which to stress these boards, with the expectation that any acoustic signals would be from events that did not cause pad cratering. It was concluded that the acoustic testing would be the most productive in the range of 100–250 g, in order to identify and validate signals of interest.

Acoustic data were measured with a DigitalWave FM-1 system using four B1025 transducers, which are approximately 9 mm in diameter. This system continuously records data to buffer, and saves a pre-set amount of waveform data whenever the signal exceeds a threshold. The sampling rate was 50 MHz, and 8192 acoustic waveform data points were saved post-trigger, while 819 points were saved pre-trigger. The overall pre-amp was set at 12, 24, or 30 dB; the recorded signal was amplified 12 or 18 dB; and the trigger was amplified 3 or 6 dB.

Acoustic data was analyzed post-test by first assessing events which are known to have occurred within the sensor array. Figure 10.3 illustrates two typical waveform types that were analyzed, showing screen captures of (a) the first set of waveforms captured on all four sensors for 225 g shock test on panel Z992 pertaining to the shock itself and (b) the fourth set of waveforms captured on all four sensors for the same test pertaining to a transient event which occurred within the sensor array. The first waveform of a given file always corresponded to the shock event itself (Fig. 10.3a) followed by other waveforms which were associated with other acoustic events resulting from the vibration of the board as well as potential damage sources in the interconnections (Fig. 10.3b). For the AE capture settings used, typical waveform capture windows

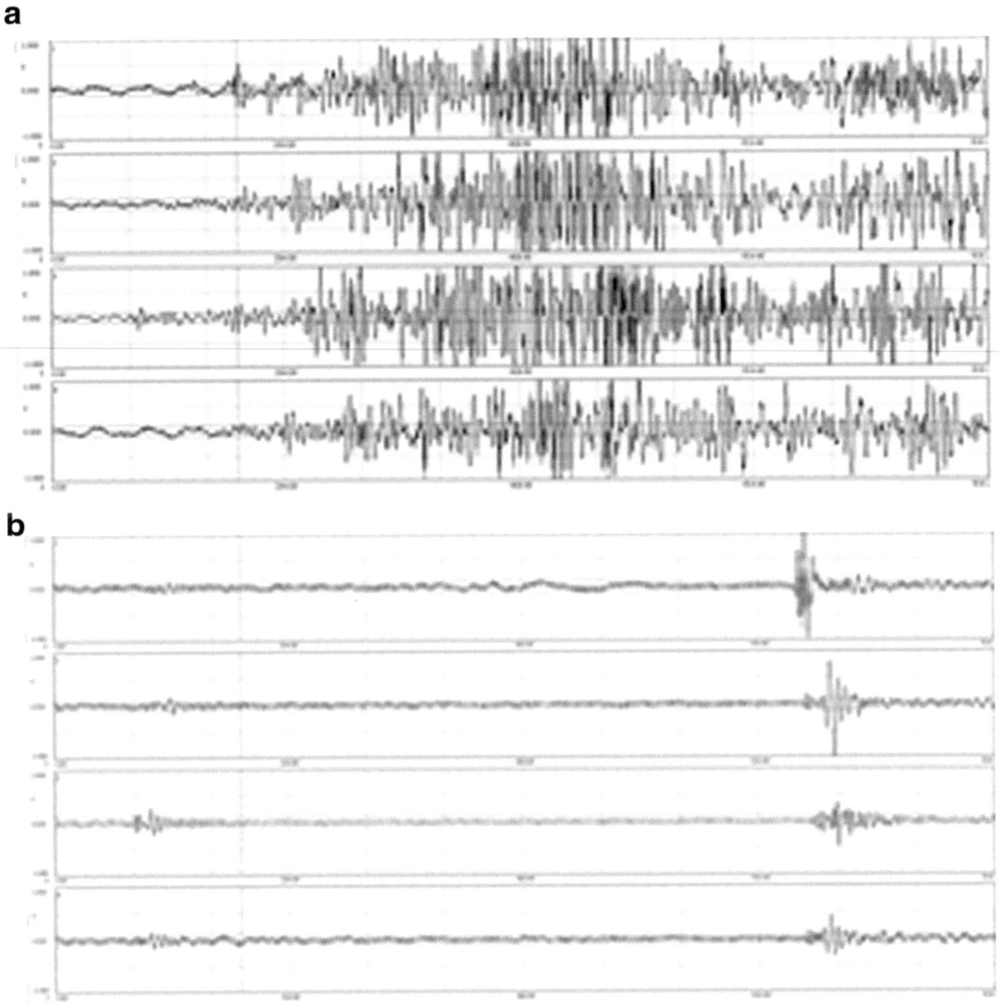

Fig. 10.3 Typical waveforms in the shock tests indicating (**a**) the shock event and (**b**) potential damage events

lasted for fairly long times (819 or 1638 microseconds) and would include several events, as in Fig. 10.3b where there is a smaller amplitude pulse in the beginning of the waveform on each sensor followed by a higher amplitude pulse on each sensor at a later time on each sensor.

Only events in which there was confidence that the event had occurred within the sensor array were analyzed. The criteria for events of interest were as follows: (1) pulses within a given waveform that clearly existed on each sensor, (2) pulses with a clear first peak which pertained to the extensional part of the waveform on at least the three sensors closest to the source and (3) pulses that had a high amplitude (>0.7 V) for the pulse with the earliest time of arrival. The clear first peak was necessary in order to estimate location and we assumed that higher amplitude events which all clearly emanated from within the sensor array were from some damage event within the board or package.

Location of events was then determined by triangulation from the times of arrival for the pulses using the following equations:

$$\sqrt{(x-x_3)^2+(y-y_3)^2}-\sqrt{(x-x_2)^2+(y-y_2)^2}=C_c(t_3-t_2) \tag{10.1}$$

$$\sqrt{(x-x_4)^2+(y-y_4)^2}-\sqrt{(x-x_2)^2+(y-y_2)^2}=C_c(t_4-t_2) \tag{10.2}$$

$$\sqrt{(x-x_1)^2+(y-y_1)^2}-\sqrt{(x-x_2)^2+(y-y_2)^2}=C_c(t_1-t_2) \quad (10.3)$$

Where x and y are the coordinates in an x–y grid, the subscript pertains to the locations of the transducers, C_c is the speed of sound in this material (3200 m/s) and t is the arrival time at a given sensor. Boron fiber (275 μm diameter, Specialty Materials, Lowell Massachusetts) breaks were used to create acoustic source events of a known location to measure the speed of sound in the material and to determine the accuracy of the triangulation method. Figure 10.4 shows sensor location (large circles) and lead-break location (small filled circles) for an undamaged board. The open circles estimate the boron fiber break location from triangulation equations. It was found that, in all but one case, the accuracy of the triangulation method was within ±5 mm of the sensor location (Fig. 10.4).

10.3 Results

The AE data were examined from six different boards that were subjected to shock tests. In some cases multiple shock tests were performed on a single board. Layout dye stain was used after the acoustic emission testing in order to validate whether acoustic events of interest were indeed the result of pad cratering events. Results are listed in Table 10.2.

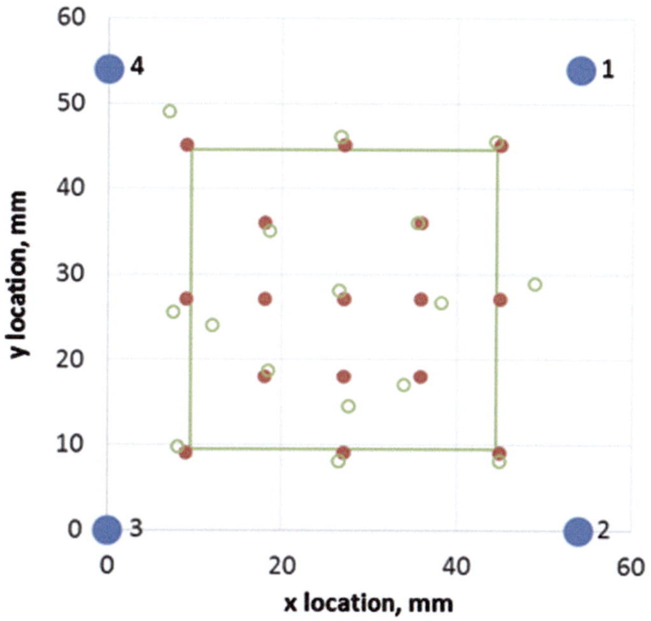

Fig. 10.4 Sensor location and lead-break location for an undamaged board

Table 10.2 Shock tests, AE results and die penetrant results

Board	Preamp (dB)	Drop inputs	FA results	Acoustic events
Z986	30	1 @ 250 g	3 partials at corner 1	Very high amplitude event near sensor 1
Z988	42	8 @ 130 g	None	Several high amplitude events around sensor 2
Z990	42	1 @ 175 g	Sensor 3 one partial	Only one high amplitude event on near 2 and lower amplitude events near 3
Z991	42	5 @ 150 g	None	One high amplitude event at 3
Z992	42	5 @ 150–225 g	1 partial sensor 1 and 2 partials sensor 3	Multiple high amplitude events near sensors 1 and 3 and a single event near sensor 2
Z997	42	10 @ 100 g	None	No high amplitude events

There are several considerations and complications in analyzing the acoustic data, including: the nature of shock tests, the difficulty in separating out signals of interest, and the inability to acquire all the pertinent acoustic data due to acquisition blackout times.

- The shock test consists of a number of mechanical events that can generate acoustic events. The shock table first contacts the base plate of the shock equipment, the table then decelerates to a stop, followed by a rebounding of the board at its natural frequency. Interconnect damage is most likely to occur as the table comes to a stop and the interconnections are loaded at maximum tension. Damage may also occur at the peak of each subsequent rebound.
- There is considerable acoustic activity from the drop due to the shock event, the sound created from the impingement of the board on the support pins, reflections of pulse waves from the edges of the board and holes in the board, and from failure sources associated with the interconnections as well as damage within the polymer composite. The damage to the interconnections are the events of interest. Sorting out the pertinent data requires good correlation between the appropriate signals and quantifiable damage.
- The computer acquires data within a fixed time window, either 1638 μs (8192 points at 5 MHz capture rate) or 819 μs (4096 points at 5 MHz capture rate) for the tests performed for this study. The computer then needs to rearm in order to save the next waveform. For the AE system used in this study it is apparent that only about 40 % of the data can be captured during the first few hundredths of a second for the 1638 μs test and only about 30 % of the data can be captured for the 819 μs test due to the blackout that occurs during the rearm time of the computer. Therefore, 60–70 % of the most critical time period could not be measured.

Due to the above concerns, only events that had clear (higher frequency) extensional waves at the first peak and relatively high peak amplitudes (>0.7 V at the transducer with the earliest arrival) were chosen for evaluation. Of the six tests in Table 10.2, there were two clear positive associations of AE events with dye stain damage (Z986 and Z992) and one clear association of no pertinent AE event with no observable damage in dye stain (Z997). The other three panels did show one or more events in regions that were not associated with damage.

The two most successful correlations with the dye stain results were for boards Z986 and Z992. Figure 10.5 shows acoustic regions of interest from the pertinent waveforms for (a) board Z986 after 275 g shock and (b) board Z992 after 225 g shock. The frequency content of these waveforms is primarily in the 100–300 kHz range. The first peaks are identified with arrows from which location analysis was performed. Note that the Z986 test had the lower preamp setting but had the higher energy pulse (several peaks >1 V). This was the only event which showed a very high signal intensity of all the events which occurred within the sensor array. All of the other events appeared more like the Fig. 10.5b example where only one or two peaks of a given pulse exhibited a high amplitude. It is believed that such a high energy event was recorded for the board in Fig. 10.5a because the longer time window settings of that board enabled 10 % more time to be monitored and the lower preamp settings did not allow low energy events to trigger the system, which may mask other high energy events within the initial critical time period.

The location analysis for the two tests are shown in Fig. 10.6. Board Z986 showed significant damage around sensor 1, which was in the region where the high energy event occurred within 0.016 μs of the initial acoustic signal associated with the shock event. The Z992 board drops exhibited events which occurred primarily on sensors 1 and 3 in the sequential shock tests with increasing acceleration. Note that one event did occur near sensor 2 but the other seven events occurred near sensors 1 and 3, which corresponded with failure analysis. These results are considered promising, but are only tentative as most of the time during the shock and reverberation portion of the test was spent re-arming the AE data acquisition equipment. It is also important to note that the 12 dB preamp setting between tests Z986 and Z992 corresponds to a difference in perceived signal amplitude. In other words, a 1 V signal at the 42 dB setting of board Z992 would correspond to about a 0.25 V reading on board Z986 at 30 dB preamp setting—which further accentuates the strength of the Z986 pulse. It is possible that the pulse from board Z986 is actually a more significant failure event and the other lower energy events associated with the other panels are not the primary fracture events, but correspond to secondary cracks, crack growth or polymer composite damage.

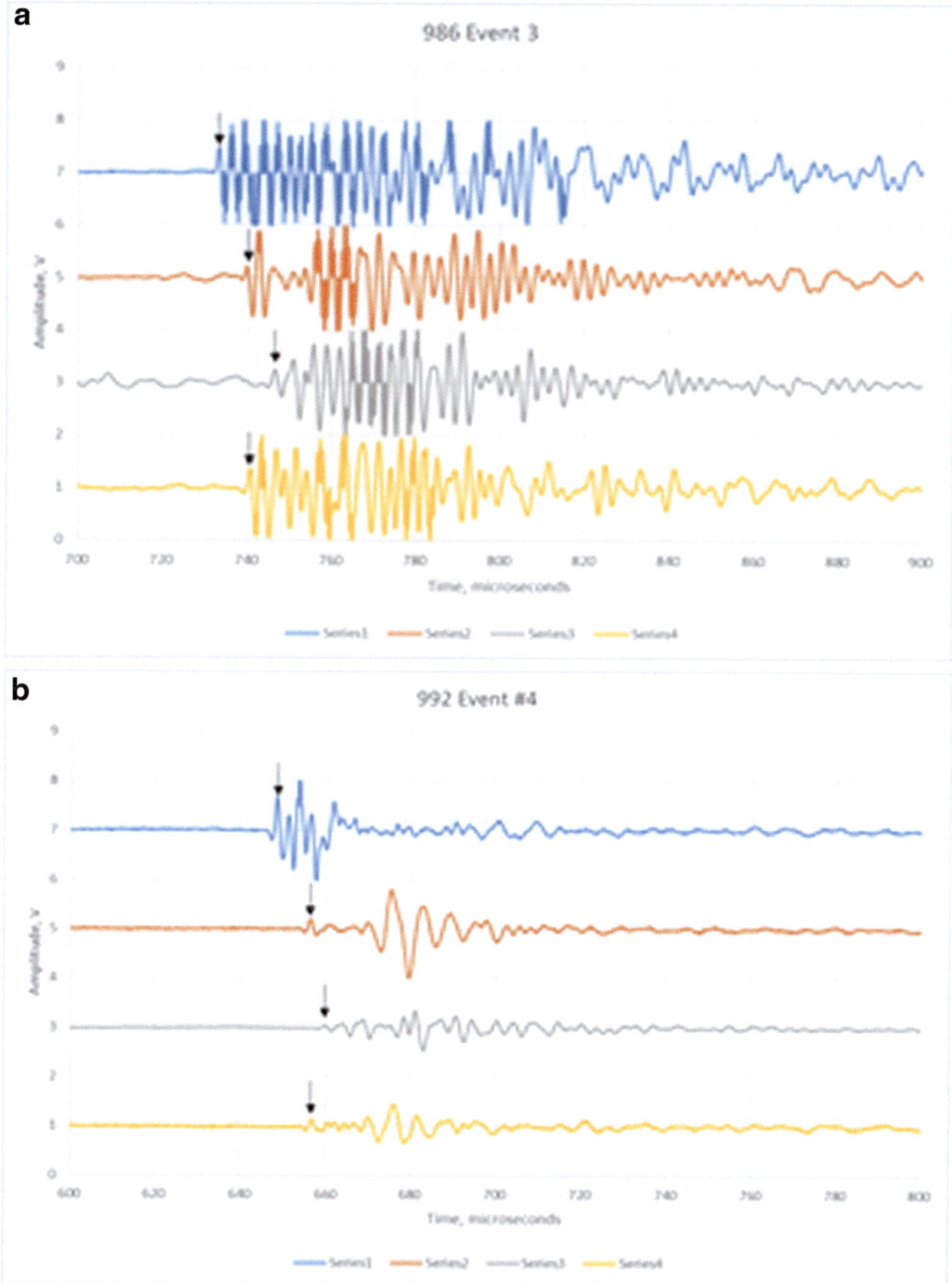

Fig. 10.5 (**a**) The high energy event from Z986 after 275 g shock and (**b**) the pulse later in the waveform

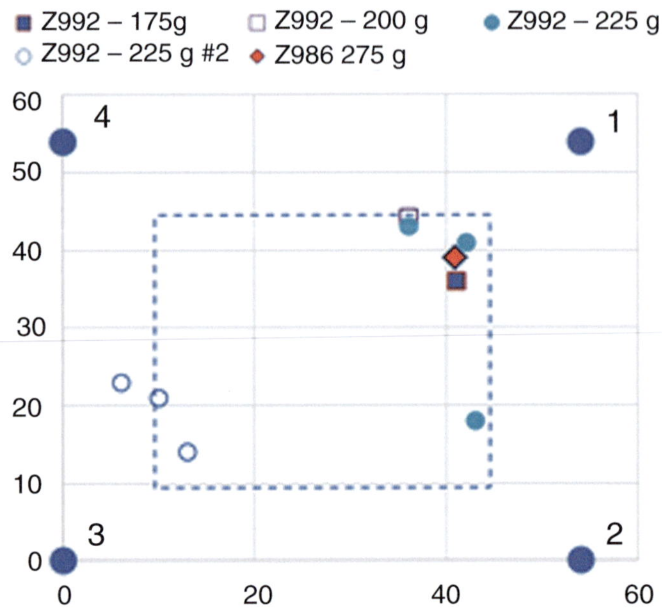

Fig. 10.6 Location analysis of pulse events from the two different shock tests for panels Z986 and Z992

10.4 Conclusions

These results demonstrate that acoustic emissions from interconnect fractures can be detected in shock. The failures that were recorded corresponded to pad craters, which is a common failure mode. The pad crater events can be distinguished from other events that are not of interest, and the locations of the pad crater events can be located within a few millimeters. Detection of the pad crater events yields the precise moment of damage.

Though the AE results show some promise, they are not conclusive. The most obvious issue is the data that is potentially lost in the first couple of hundredths of a second after the initial shock event. Therefore, the most significant change to the test procedure would be to take a streaming approach to acquiring data over very long time windows so that the entire test time could be observed. This is possible with the current acoustic data acquisition system by increasing the number of points and decreasing the capture rate. For example, increasing the points to 32,768 (the highest available on the system) and decreasing the capture rate to 2 MHz would result in a time window of 0.016 s while still maintaining a frequency resolution of 250 kHz. A high fidelity digital oscilloscope would also be a viable option.

If the entire time window is captured, and pertinent events can be identified, then one could determine appropriate AE settings that would better isolate individual events of interest using the more common "hit-based" approach of AE.

References

1. Bansal, A: A new approach for early detection of PCB pad cratering failures. In: IPC APEX EXPO proceedings, Las Vegas, NV. (2011)
2. Bansal, A., Guirguis, C., Liu, K.C.: Investigation of Pad Cratering in large Flip-Chip BGA using acoustic emission. In: IPC APEX EXPO Proceedings, San Diego, CA. (2012)
3. Ralph, W., Daspit, G., Cain, A.W., Benedetto, E.E., Jenkins, R.S., Allen, A.M., Newman, K.: Acoustic emission detection of BGA components in spherical bend. In: Proceedings of the Electronic Components and Technology Conference, Las Vegas, NV, pp. 208–213. (2013)
4. Ralph, W., Daspit, G., Cain, A.W., Benedetto, E.E., Jenkins, R.S., Allen, A.M., Newman, K.: Acoustic waveform energy as an interconnect damage indicator. In: Conference Proceedings of the Society for Experimental Mechanics, pp. 33–42. (2014).
5. Ralph, W., Benedetto, E.E., Allen, A.M., Newman, K.: Pad crater detection using acoustic waveform analysis. In: Proceedings of the Electronic Components and Technology Conference, Orlando, FL, pp. 1433–1440. (2014)

Chapter 11
Application of Nanoindentation and Microdiffraction to Study Aging Effects in Lead Free Solder Interconnects

Md Hasnine, Jeffrey C. Suhling, Barton C. Prorok, Michael J. Bozack, and Pradeep Lall

Abstract In this work, we have performed a study on aging effects in actual solder joints extracted from PBGA assemblies using nanoindentation. The tested joints were extracted from 14 × 14 mm PBGA assemblies (0.8 mm ball pitch, 0.46 mm ball diameter) that are part of the iNEMI Characterization of Pb-Free Alloy Alternatives Project (16 different solder joint alloys available). Using nanoindentation techniques, the stress-strain and creep behavior of the aged SAC solder materials were explored at the joint scale for various aging conditions. After aging, the joints were loaded in the nano-indentor, and the load-deformation behavior during indentation has been used to characterize the mechanical properties of the solder joint for various aging conditions including modulus, hardness, and yield stress. Using constant force at max indentation, we have also measured the creep response of the aged and non-aged solder joint materials for various stress levels. With this approach, we have been able to quantify aging effects in joints and correlate their magnitudes to those observed in tensile testing of miniature bulk specimens. Our results have shown that the aging induced degradations of the mechanical properties (modulus, hardness) of single grain SAC305 joints were similar to those seen previously by testing of larger "bulk" solder specimens. However, the degradation of the creep response, while still significant, was less in the solder joints relative to larger uniaxial tensile specimens. This was due to the single grain nature of the joints considered, and the lack of the grain boundary sliding creep mechanism.

We have recently extended our work in the nanoindentation area to examine a full test matrix of solder alloys. In particular, we have characterized the effects of silver content on SAC solder aging by testing joints from SAC105, SAC205, SAC305, and SAC405 test boards assembled with the same reflow profile. Our results show that the aging induced mechanical behavior degradations in joints are similar to those found observed in very small tensile specimens with 10–20 grains. These effects are strongest in SAC alloys with low silver content (e.g. SAC105). The observed mechanical behavior changes in joints are due to evolution in the microstructure and residual strains/stresses in the solder material, and measurements of these evolutions are critical to developing a fundamental understanding of solder joint aging phenomena. As another part of this work, we have performed an initial study of these effects in the same SAC305 solder joints that were tested using nanoindentation. The enhanced x-ray microdiffraction technique at the Advanced Light Source (Synchrotron) at the Lawrence Berkeley National Laboratory was employed to characterize several joints after various aging exposures (0, 1 and 7 days of aging at $T = 125$ °C). For each joint, microdiffraction was used to examine grain growth, grain rotation, sub-grain formation, and residual strain and stress evolution as a function of the aging exposure. The entire joints were scanned using a 1 micron step size, and the results were correlated with changes in the mechanical response of the joint specimens measured by nanoindentation.

Keywords: Nanoindentation • Lead Free Solder • Microdiffraction • Modulus • Hardness • Creep

11.1 Introduction

Solder joints are an important part of microelectronic packages because they serve as mechanical as well as electrical interconnects. However, environmental and health concerns over the use of lead in microelectronic packaging have led to the vigorous development of alternative solder alloys largely based on Sn. Among several lead free solders, Sn-Ag-Cu (SAC) alloys have been identified as potential candidates for the replacement of standard Sn-Pb solder for ball grid array

M. Hasnine • J.C. Suhling (✉) • B.C. Prorok • M.J. Bozack • P. Lall
Department of Mechanical Engineering, Auburn University, Auburn, AL 36849, USA

Center for Advanced Vehicle and Extreme Environment Electronics (CAVE3), Auburn University, Auburn, AL 36849, USA
e-mail: jsuhling@auburn.edu

B.C. Prorok, L. Starman (eds.), *MEMS and Nanotechnology, Volume 5*,
Conference Proceedings of the Society for Experimental Mechanics Series, DOI 10.1007/978-3-319-22458-9_11

application. For enhanced reliability of portable electronic devices during accidental shock loading (e.g. high strain rates) several alloys with low silver content have been recommended.

Thermal fatigue and creep performance of solder joints are critically important for high reliability applications where they are exposed to isothermal aging. Solder fatigue life degrades faster as joints are subjected to aging [1]. The mechanical properties of a lead free solder are strongly influenced by its microstructure, which is controlled by its thermal history including solidification rate and thermal aging after solidification. Such aging effects are greatly exacerbated at higher temperatures typical of thermal cycling qualification tests. Aging degrades the solder joint mechanical properties under elevated temperature service conditions. However, significant changes occur even with aging at room temperature.

In our prior papers on elevated temperature aging effects [2–14], we have demonstrated that the observed material behavior variations of SACN05 (N = 1, 2, 3, 4) lead free solders during isothermal aging at a variety of elevated temperatures (e.g. 25, 50, 75, 100, and 125 °C) were unexpectedly large and universally detrimental to reliability. The measured stress-strain data demonstrated large reductions in stiffness, yield stress, ultimate strength, and strain to failure (up to 50 %) during the first 6 months after reflow solidification. After approximately 10–20 days of aging, the lead free solder joint material properties were observed to degrade at a slow but constant rate. In addition, even more dramatic evolution was observed in the creep response of aged solders, where up to 10,000× increases in the secondary creep rates were observed for aging up to 6 months. The aged solder materials were also found to enter the tertiary creep range (imminent failure) at much lower strain levels than virgin solders (non-aged, tested immediately after reflow solidification).

There have been limited prior mechanical loading studies on aging effects in actual solder joints extracted from area array assemblies (e.g. PBGA or flip chip) [15–18]. This is due to the extremely small size of the individual joints, and the difficulty in gripping them and applying controlled loadings (tension, compression, or shear). Pang, et al. [15] have measured microstructure changes, intermetallic layer growth, and shear strength degradation in custom SAC single ball joint lap shear specimens subjected to elevated temperature aging. Darveaux [16] performed an extensive experimental study on the stress-strain and creep behavior of solder using specially constructed double lap shear specimens with a 10 × 10 area array solder balls. He found that aging for 1 day at 125 °C caused significant effects on the stress-strain and creep behavior. For example, aged specimens were found to creep much faster than non-aged specimens by a factor of up to 20 times for both SAC305 and SAC405 solder alloys. Wiese, et al. [17] also studied the effects of aging on solder joint creep using custom assemblies with four flip chip solder balls, and found highly accelerated creep rates after aging at 125 °C. Finally, Dutta and coworkers [18] used an impression creep technique with a cylindrical punch to study creep in PBGA solder balls that had been subjected to thermal-mechanical cycling.

Nanoindentation techniques have been widely used to probe the mechanical properties and deformation behavior of extremely small material samples [19]. Over the past decade, it has been applied by several investigators to characterize lead free solder joints and intermetallic compounds (IMC) in lead free solder joints [20–35]. In early studies by Rhee, et al. [20], Chromik, et al. [21], and Deng and coworkers [23, 24], the elastic modulus E and hardness H of various regions in Sn-Ag and SAC lead free solder joints were explored by nanoindentation. In particular, the properties for the Sn-rich phase (β-tin) and the eutectic phase (containing β-tin and a mix of Sn-Ag and Sn-Cu intermetallics) were explored. Attempts were also made to indent individual Sn-Ag and Sn-Cu intermetallic particles [27, 29, 33], and then to compare the properties of the IMCs to those for the solder joint phases. Rhee, et al. [20] also measured changes in the mechanical properties after the joints were subjected to thermal-mechanical cycling.

As discussed above, there is an extensive literature that documents the large changes in the microstructure and mechanical behavior that occur in bulk lead free solder specimens during isothermal aging. There have also been some limited investigations on the effects of aging on mechanical properties and creep behavior of solder joint arrays. In addition, nanoindentation has been utilized to study aging induced changes in the mechanical properties of intermetallic compound layers in solder joints. However, there has been little work on aging effects on mechanical properties and creep behavior in individual solder joints. Such knowledge is crucial for the optimizing the design, manufacturing, and reliability of microelectronic packages. Characterization of individual joints is quite challenging because of their extremely small size, and the difficulty in gripping them and applying controlled loadings.

In our ongoing research, we have extended our previous work on nanoindentation of joints to examine a full test matrix of SACN05 solder alloys extracted from PBGA assemblies. The effects of silver content on SAC solder aging has been evaluated by testing joints from SACN05 (SAC105, SAC205, SAC305, and SAC405) test boards assembled with the same reflow profile. In all cases, the tested joints were extracted from 14 × 14 mm PBGA assemblies (0.8 mm ball pitch, 0.46 mm ball diameter) that are part of the iNEMI Characterization of Pb-Free Alloy Alternatives Project (16 different solder joint alloys available). After extraction, the joints were subjected to various aging conditions (0–12 months of aging at $T = 125$ °C), and then tested via nanoindentation techniques to evaluate the stress-strain and creep behavior of the four aged SAC solder alloy materials at the joint scale. The observed aging effects in the SACN05 solder joints have been quantified and correlated with the magnitudes observed in stress-strain and creep testing of miniature bulk specimens performed in prior studies, as well as with new test results for tests performed on very small tensile specimens with approximately ten grains. In a related study, we have used the x-ray microdiffraction beamline at the Advanced Light Source (Synchrotron) at the Lawrence Berkeley National Laboratory to study the evolution in the microstructure and residual strains/stresses in solder joint materials after various aging exposures (0, 1, and 7 days of aging at $T = 125\ ^\circ C$). For each joint, microdiffraction was used to examine grain growth, grain rotation, subgrain formation, and residual strain and stress evolution as a function of the aging exposure. Rectangular regions at the centers of joints were scanned using a 1×1 micron step size, and the results were correlated with changes in the mechanical response of the joint specimens measured by nanoindentation.

11.2 Experimental Procedure

11.2.1 Solder Joint Samples

The lead free solder joints in this study were extracted from PBGA assemblies (Amkor CABGA, 14×14 mm, 192 balls, 0.8 mm ball pitch, 0.46 mm ball diameter). The test boards were assembled as part of the iNEMI Characterization of Pb-Free Alloy Alternatives Project [36], and a variety of samples with 16 different solder ball alloys are being studied. In this work, the BGA ball alloys considered were SAC105 (98.5Sn-1.0Ag-0.5Cu), SAC205 (97.5Sn-2.0Ag-0.5Cu), SAC305 (96.5Sn-3.0Ag-0.5Cu), and SAC405 (95.5Sn-4.0Ag-0.5Cu). These four alloys have been collectively referred to as SACN05 in this paper, with N representing the integer percentage value (1, 2, 3, and 4) of the silver content of the ball alloy of the unassembled BGA component. The test boards had ENIG surface finish, and SAC305 solder paste was used in the surface mount assembly process for all of the ball alloys. The assembled PBGA components were cut out from the test boards and then cross-sectioned into samples of four joints each. These samples were mounted in an epoxy molding compound suitable for SEM microscopy, and then polished to a level appropriate for nanoindentation. Details of the sample preparation process include mechanical grinding with several SiC papers (#320 to #400, #600, #800 and #1200), and then final polishing with 1 μm diamond paste followed by 0.05 μm colloidal silica suspensions. This resulted in mirror finish samples suitable for nanoindentation tests and SEM microscopy.

11.2.2 Nanoindentation System and Test Procedures

The nanoindentation tests were conducted using an instrumented MTS Nanoindenter XP system with a Berkovich tip indenter [37–38]. In addition, the Continuous Stiffness Measurement (CSM) technique [39] was also used in all experiments to extract elastic modulus and hardness as a function of indentation depth. A typical cross-sectioned lead free solder ball sample after nanoindentation testing is shown in Fig. 11.1, and a close-up view of an example permanent indentation mark is shown in Fig. 11.2. All tests were performed on single grain (Sn crystal) solder balls, so that there were no orientation effects caused by an indentation array covering two or more grain boundaries of grains with significantly different crystal orientation (different material properties).

A maximum load of 30 mN was selected for the tests so that the indentation marks were large enough to cover all the phases of SAC305 solder material (see Fig. 11.2). Thus, the nanoindentation tests characterized the global mechanical properties of the solder joints, rather than the localized properties in the β-Sn phases (Sn-rich dendrites) or in the eutectic phases between dendrites that contain Sn-Ag and Sn-Cu intermetallics. Indentation experiments were conducted at constant

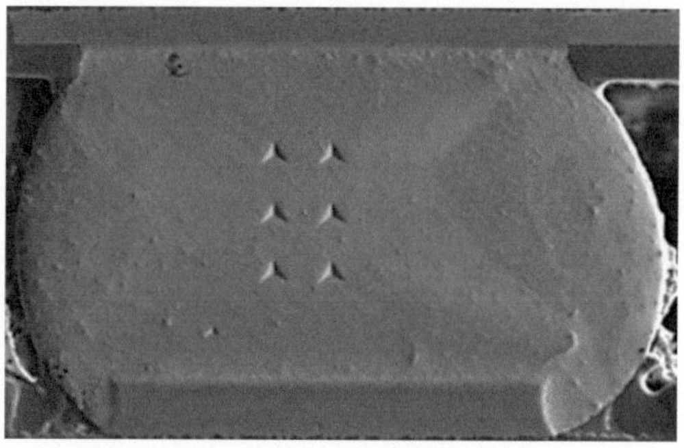

Fig. 11.1 Solder ball after nanoindentation testing

Fig. 11.2 Permanent indentation after testing

indentation displacement rate of 10 nm/s, corresponding to an effective strain rate of 0.05 s^{-1}. Calibration of the load and hardness measurements was performed on fused silica.

Indentation creep tests were performed holding the load constant at its maximum value of 30 mN for 500–1000 s. After this dwell, the sample was unloaded at the same 10 nm/s displacement rate used during loading. There was also a holding period of 60 s after 90 % unloading to allow for thermal drift correction. In our work, the thermal drifts were kept smaller than 0.05 nm/s, and the thermal drift effects were excluded from the resulting displacement data. The details of measurement and processing of nanoindentation data for lead free solders are discussed in detail in our earlier paper [13].

11.2.3 Aging Studies on SACN05 Solder Joints

Aging phenomena in lead free solder joints have been explored by nanoindentation testing of several SAC samples exposed to various aging conditions prior to testing. The SAC alloys considered were from BGA components fabricated using four different SACN05 ball alloys (SAC105, SAC205, SAC305, and SAC405). Using the nanoindentation methods detailed above, the stress-strain and creep behavior of the SAC solder joints have been explored for various aging conditions. Mechanical properties characterized as a function of aging include the elastic modulus and hardness, as well as the creep strain rate vs. applied stress response. With these approaches, aging effects in solder joints were quantified and correlated to the magnitudes of those observed in stress-strain and creep testing of miniature bulk specimens with hundreds of grains performed in prior studies [2–9]. In addition, new creep experiments with aging have been performed on very small tensile specimens with approximately ten grains, and these data have also been correlated with the nanoindentation results.

Lead free solder joints exhibits highly anisotropic characteristics in its mechanical, thermal, and diffusion properties. For example, the modulus of elasticity of tin in the (001) direction (67.6 GPa) is three times higher than that in the (100) direction (23.6 GPa) [40]. Single BGA solder joints have been observed to typically have 1–6 grains after reflow [35–37]. The Sn grain/crystal orientations and the intermetallic phases present in a solder joint will play important roles in determining its mechanical behavior and the reliability. Due to the highly anisotropic nature of Sn crystals, a variety of mechanical properties can be found in a single joint that contains multiple grains. We have limited our testing to single grain solder joints. Thus, it is important to understand the precise grain orientation and the crystallographic direction normal to the polished surface in each tested joint so that results from multiple joints can be compared. Polarized light microscopy and Electron Back Scattered Diffraction (EBSD) techniques have been utilized for this purpose.

11.2.4 Experimental Test Matrix

The aging induced changes in mechanical behavior of single crystals solder joints have been examined included four SAC105 joints, three SAC205 joints, two SAC305 joints, and three SAC405 joints. For each joint, different sets of aging conditions were explored. The aging conditions completed to date include: no aging; and 10, 30, 90 and 180 days of aging at $T = 125\ ^\circ C$. For each aging condition, a 2×3 array of indentations was performed on each ball. The samples were stored in a freezer at $T = -10\ ^\circ C$ prior to cross-sectioning to minimize any aging effects after board assembly. Aging of the epoxy mounted samples was then performed in a box oven, and light polishing was performed to remove any oxides after each oven exposure. For each aging condition and solder joint, the individual and average results for the 6 indentations have been reported for the elastic modulus, hardness, and creep strain rate vs. stress curves. Additional testing has been performed on very small bulk specimens (SAC105 and SAC305) to correlate the uniaxial creep with the nanoindentation creep. The uniaxial test sample preparation procedure has been described in references [2–8].

11.2.5 Solder Joint Crystal Orientations

Using polarized light optical microscopy on an Olympus BX60 metallographic microscope system, it was verified that each of the tested SACN05 solder joints were comprised of a single grain. The unique colors and contrasts obtained for the samples under polarized light also indicated that the crystal orientations were different. To more rigorously understand the joint orientations, Electron Backscatter Diffraction (EBSD) was performed [41–43]. A scanning map with a step size 5 μm in each direction was obtained for each joint using an EBSD system (Oxford HKL Channel 5) attached to a JEOL JSM-7000 F SEM. For each scanning point, the Electron Backscatter Diffraction Pattern (EBSP) or so-called Kikuchi bands were indexed, and an available phase database for β-Sn was used to find the three Euler angles relative to the crystallographic directions of the tetragonal crystal. The results showed that the crystal orientation was constant across each joint. The Inverse Pole Figure (IPF) for the two SAC 305 joints are plotted in Fig. 11.3.

Analogous Inverse Pole Figure (IPF) plots for the tested SAC105, SAC205, and SAC405 solder joints are shown in Figs. 11.4 and 11.5. For each alloy, the joint orientations were unique for each joint in the set. Using the modulus map in Fig. 11.6, it is expected that joint #1 will have the largest modulus and hardness for the SAC105 joints. Likewise, joint #1 is expected to have the largest modulus and hardness for the SAC205 joints, and joint #2 is expected to have the largest modulus and hardness for the SAC405 joints.

Fig. 11.3 Inverse pole figures for solder joints (SAC305)

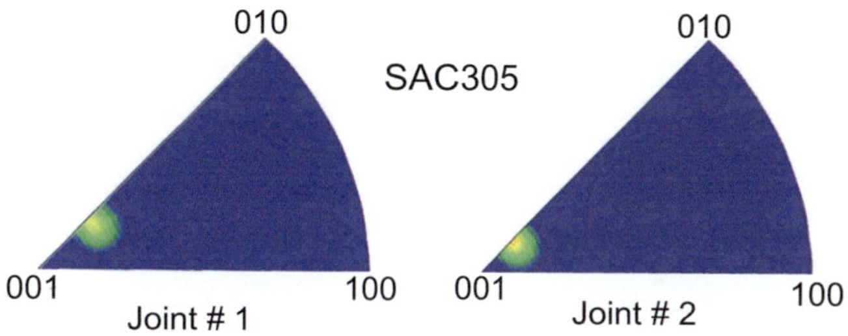

Fig. 11.4 Inverse pole figures for solder joints (SAC105)

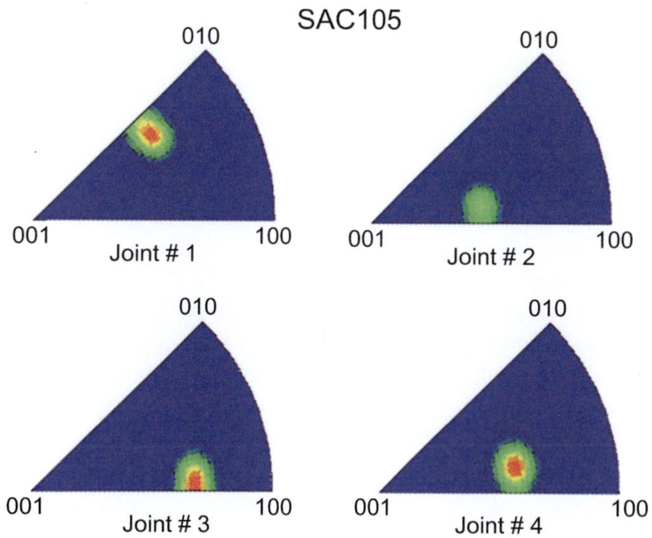

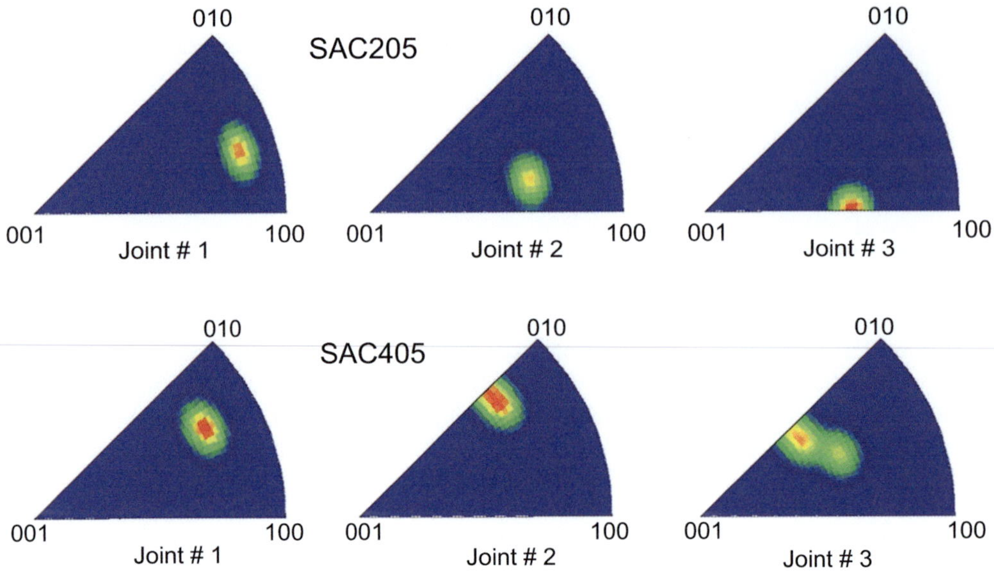

Fig. 11.5 Inverse pole figures for solder joints (SAC205 and SAC405)

Fig. 11.6 Elastic modulus of β-Sn crystal as a of orientation [40]

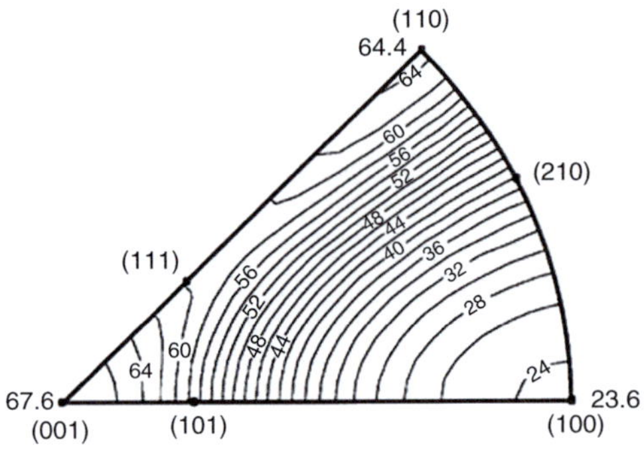

11.3 Results

11.3.1 Effect of Aging on Elastic Modulus and Hardness

The CSM method was used to measure elastic modulus and hardness values for each aging condition on each solder joint. For the case of no aging, the effective modulus of the four SAC105 joints varied from 50.66 GPa to 42.54 GPa for the different orientations, which demonstrates the strong crystal anisotropy of the solder joints. It was also observed that for 180 days of aging at 125 °C, that the effective modulus of the four SAC105 joints #1, #2, #3 and #4 degraded by 22.72 %, 30.62 %, 31.38 %, and 32.92 %, respectively. The corresponding degradations of the hardness over the same period were 39.80 %, 42.06 %, 43.12 %, and 45.92 %. These percentage reductions with 180 days of aging at 125 °C are very similar to those observed for the elastic modulus (51 %) and yield stress (43.32 %) of SAC105 measured by tensile testing of larger multi-grain uniaxial samples in our prior aging investigations [4–6]. The uniaxial samples from prior investigations have been examined using polarized light microscopy, and are known to contain hundreds of grains with a variety of orientations.

Plots of the mechanical property evolution with aging time for the SAC105 joints are presented in Figs. 11.7 and 11.8. The evolutions of the mechanical properties with aging were well fit using linear and exponential decay empirical models:

$$E = C_0 + C_1 t + C_2 e^{-C_3 t} \tag{11.1}$$

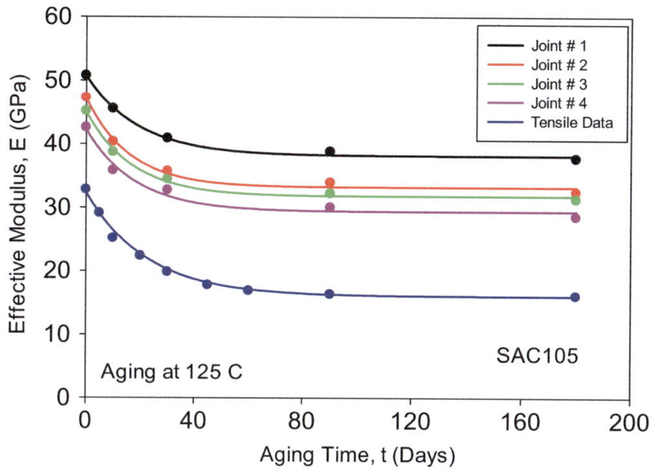

Fig. 11.7 Evolution of modulus with aging time (SAC105, aging at 125 °C)

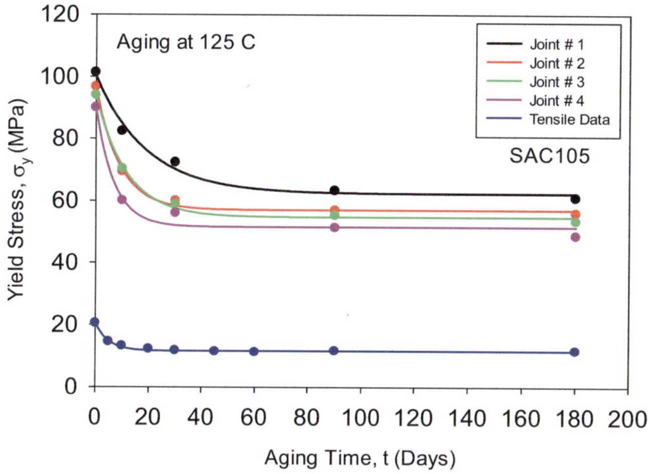

Fig. 11.8 Evolution of hardness with aging time (SAC105, aging at 125 °C)

$$H = C_0 + C_1 t + C_2 e^{-C_3 t} \tag{11.2}$$

where C_0, C_1, C_2, C_3 are regression constants. The majority of the property degradations in the solder joints occurred within the first 20–30 days of aging in an exponential manner. After this point, the variations are small and can be fit with a linear decay. These observations are in agreement with our aging results for testing of larger solder tensile specimens [2–6]. In addition to the nanoindentation results from the current paper, we have added our prior tensile testing results (blue curve) to the elastic modulus and yield stress data in Figs. 11.7 and 11.8. We have used Tabor hardness formula [44] to estimate the yield stress data from nanoindentation test. It is observed that the modulus and yield stress variations are qualitatively similar in shape. The percentage reductions of modulus are 25–30 % after 180 days of aging at 125 °C for both the solder joint nanoindentation and miniature bulk tensile test results, whereas the yield stress reductions are ~43 % as discussed above. However, the magnitudes of the modulus and yield stress at the joint-scale are significantly larger than those of the tensile samples for all aging times. This is not unexpected due to the single grain nature of the solder joints tested. As discussed above, the four tested SAC105 joints both had their out-of-plane orientations between the (101) and (100) direction of β-Sn. This orientation has a modulus of 67.6 GPa, which is at the upper end of the range ($23.6 < E < 67.6$) of values possible when considering all crystal orientations. We have established that the miniature bulk tensile specimens (80–100 mm in length) are typically composed to hundreds of grains with a variety of orientations. Thus, the modulus of the tensile specimens will be significantly smaller and highly averaged, on the order the average/mid-point of the modulus spectrum for β-Sn.

Similar results were obtained for aging induced degradations of the solder joints for the three other solder alloys. It is well understood from prior investigations that lowering the silver content of SAC alloy tends to decrease the stiffness and strength, and increase the creep strain rate. Our aging studies using large tensile specimens with hundreds of grains also have suggested the alloys with lower silver content are more susceptible to aging induced degradations in their mechanical properties and increases in their creep strain rates [4, 5]. The same conclusions can be made qualitatively using the aging data for single crystal SACN05 solder joints presented above. However, it is difficult to make precise quantitative comparisons of joints for the four alloys because the properties are also strongly dependent on the crystal orientation. As seen from Figs. 11.3, 11.4, and 11.5, there does not exist a set of four tested joints, one from each alloy, that have similar crystal orientations. Thus, it is difficult to separate the relative performance of the joints from different alloys.

In the discussion below, we have compared the relative changes due to aging for four joints, one from each SACN05 alloy. Aging data for up to 180 days at $T = 125\ ^\circ\mathrm{C}$ have been analyzed. The particular joints considered were SAC105 joint #2, SAC205 joint #2, SAC305 joint #3 (new), and SAC405 joint #1. Figure 11.9 illustrates a plot of the percentage reduction in the effective modulus occurring in the four joints as a function of the aging time:

$$\text{Reduction in Modulus }(\%) = \left|\frac{E_{\text{Aged}} - E_{\text{No Aging}}}{E_{\text{No Aging}}}\right| \times 100 \tag{11.3}$$

Likewise, Fig. 11.10 illustrates a plot of the percentage reduction in the hardness occurring in the four joints as a function of the aging time:

$$\text{Reduction in Hardness }(\%) = \left|\frac{H_{\text{Aged}} - H_{\text{No Aging}}}{H_{\text{No Aging}}}\right| \times 100 \tag{11.4}$$

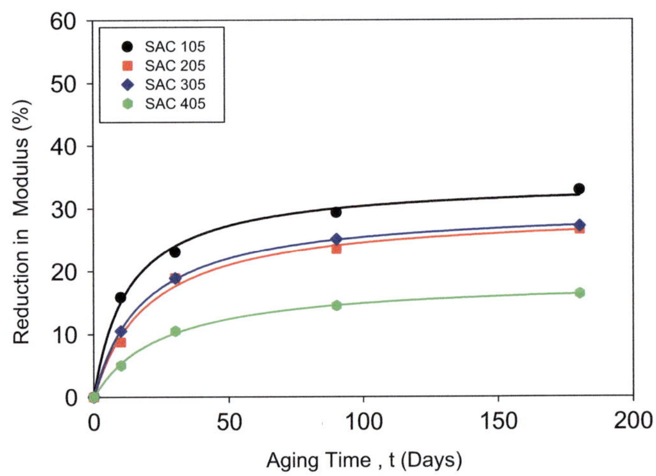

Fig. 11.9 Effect of silver content on SACN05 joint effective modulus change

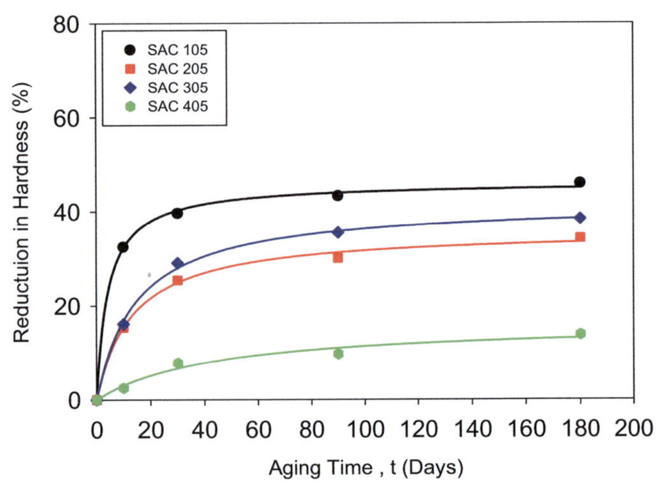

Fig. 11.10 Effect of silver content on SACN05 joint hardness change

The joint data presented in these graphs suggest that lowering the silver content tends to increase the normalized reduction in the mechanical properties (elastic modulus and hardness). The highest percentage changes (degradations) were observed for the SAC105 joint for all aging times, while the lowest changes (degradations) were seen in the SAC405 joints for all aging times.

11.3.2 Effect of Aging on Creep Response

As creep deformations are often large in solder joints in microelectronic components, it is important to investigate the degradations of the creep properties of SAC solder joints due to isothermal aging. The creep response was evaluated for various aging conditions using nanoindentation. For all of the tests, a constant peak load of 30 mN and a dwell time of 900 s were used during the hold periods. Using the recorded creep displacement vs. time data, the creep strain rate vs. applied stress curves were generated as outlined in our prior work [13].

Figure 11.11 contains the strain rate vs. applied stress responses for SAC105 joints #1, #2, #3 and #4. Each curve in these plots represents a fit of the exponential model to the nanoindentation creep data as outlined in reference [13]. Since the nanoindentation compressive stress levels during the creep experiments were typically above 50 MPa, these results are extrapolations for the stress levels below this level. The red curves in these graphs are for the no aging case, and represent the best (lowest) creep strain rates occurring in the joints. As aging progressed, the creep strain rates became much larger in each joint for every stress level, shifting the creep response curves significantly. The largest changes occurred in the first 30 days of aging, with subsequent changes occurring at slower rates. The red, blue, green, black and pink data points in each plot are creep strain rates measured by separate tensile testing of extremely small uniaxial specimens that were aged for 0, 10, 30, 90

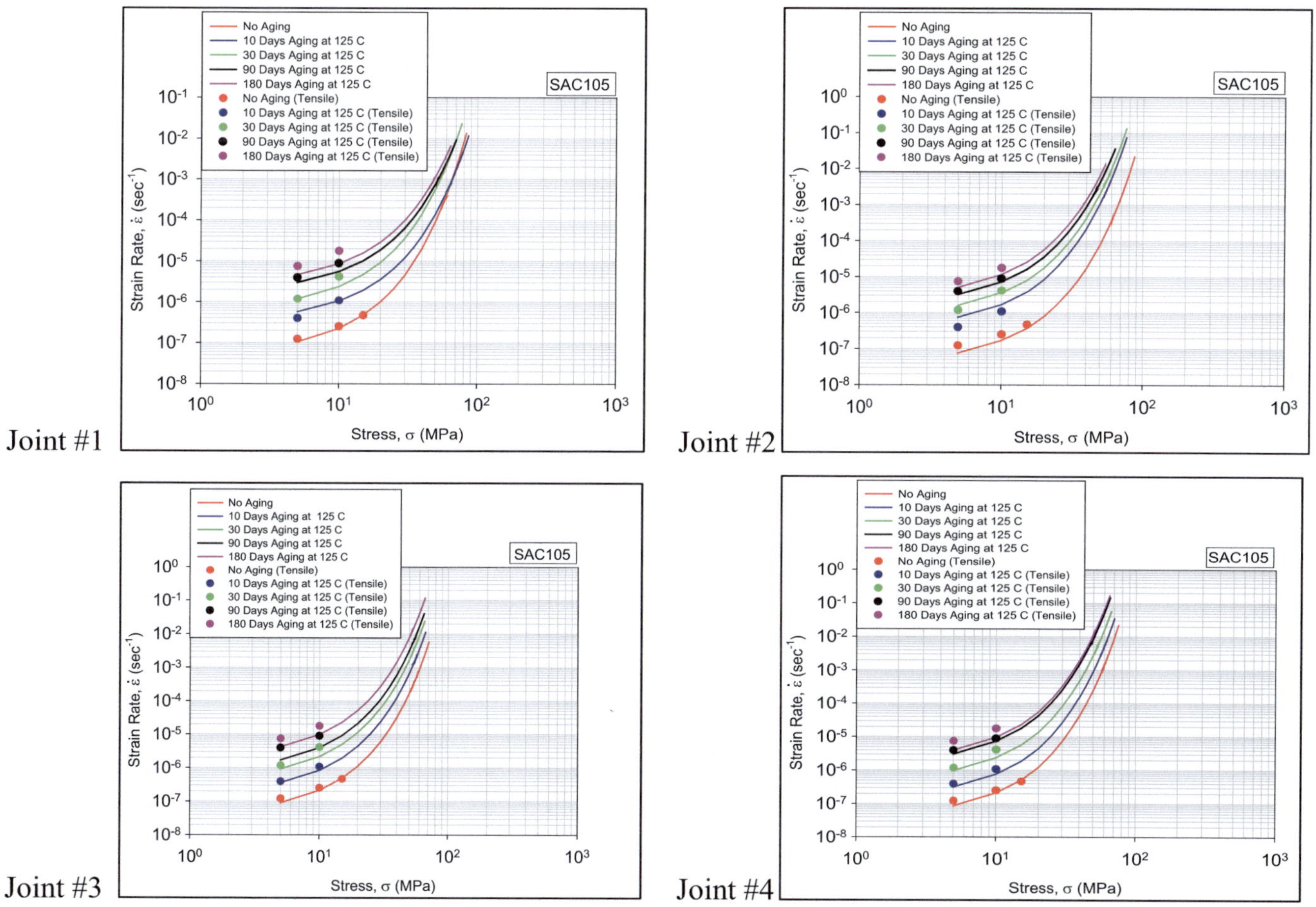

Fig. 11.11 Creep strain rate vs. stress plots (SAC105, Joints #1, 2, 3, and 4)

and 180 days, respectively. These specimens had a gage length of 10 mm, and polarized light microscopy revealed that they were composed of just a few grains. The agreement of the creep curves from nanoindentation with the small-scale tensile creep data is very good at the lower stress levels, suggesting that proposed extrapolation procedure has the potential to yield accurate predictions.

Numerical values of the creep strain rates have been extracted from Fig. 11.11 for an applied stress level of 15 MPa and plotted in Fig. 11.12 along with tensile creep data from small-scale 10 mm uniaxial samples (data points in Fig. 11.11). In addition, the creep rates measured in prior studies [5, 6] by tensile testing of longer (80–100 mm) miniature bulk uniaxial samples are also included. The increases in the creep stain rates observed for the four joint samples were very similar to those measured for the small 10 mm tensile specimens. For example, after 30 days of aging the four joints had increases of 8–19×, while the creep rate of the small tensile specimens increased by 41.38×. However, dramatically different results were obtained for the longer miniature bulk tensile samples, where an increase in the creep rate of 3200× was observed. Thus, there were significantly more aging induced degradations in the larger bulk solder tensile samples relative to the joint samples and small tensile specimens. As discussed below, this can be explained from the single grain/crystal nature of the joint samples and the small number of grains (<10) present in the small-scale tensile specimens that were tested.

Similar results were obtained for the aging induced degradations of the creep responses of the three other SACN05 solder alloys. An analogous comparison was made in Fig. 11.13 for the degradations of the creep responses of four joints of different alloys with aging. In this case, the increase in the creep strain rate was plotted as a function of aging time:

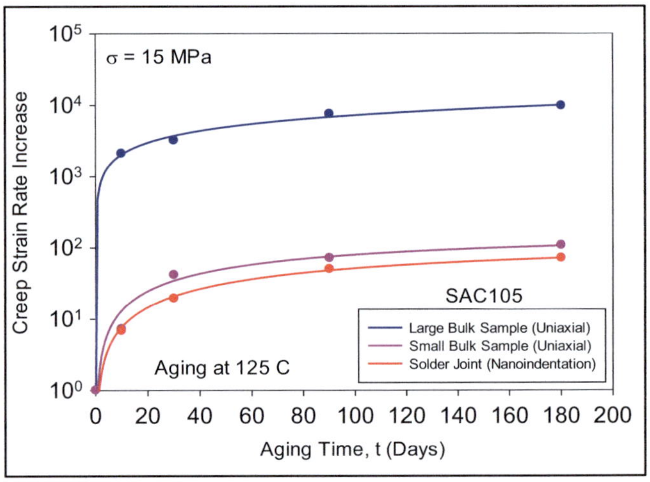

Fig. 11.12 Creep rate changes with aging (SAC105)

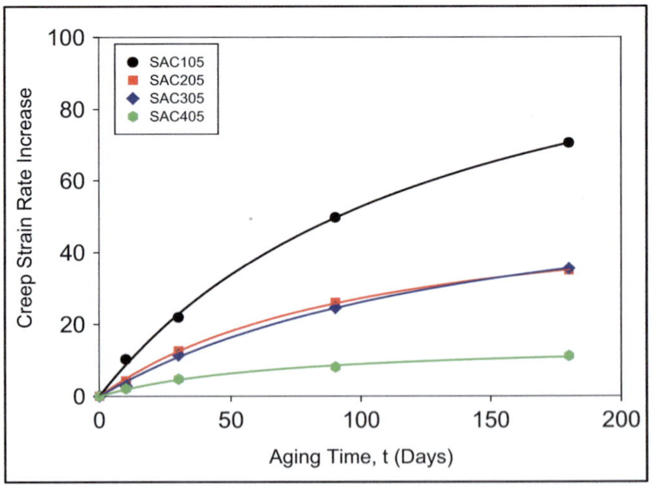

Fig. 11.13 Effect of silver content on SACN05 joint creep strain rate change (Aging at 125 °C)

$$\text{Creep Rate Increase} = \frac{\dot{\varepsilon}_{\text{Aged}}}{\dot{\varepsilon}_{\text{No Aging}}} \tag{11.5}$$

As observed previously, the largest creep rate increase (50×) was found in SAC105 (joint #2), while the smallest creep increase (8×) was found in SAC405 (joint #1). The presence of more Ag content in the SACN05 alloy gives additional precipitates in the form of Ag_3Sn IMC particles, which are responsible for hardening of Sn matrix. In addition, the lower silver content alloys (e.g. SAC105) showed larger coarsening of Ag_3Sn particles relative to the higher silver content alloys (e.g. SAC405).

11.4 X-Ray Microdiffraction

Synchrotron X-ray microdiffraction is a new powerful experimental technique available for studying mechanical properties of materials at small length scales, in the range of grains and defect interactions [45]. The method is capable of measuring the variation in internal (residual) stress, strain, grain orientation, grain rotation, and plastic deformations between grains and within individual grains. This can provide important information on changes occurring within a solder joint at small scales. The technique is based on very small submicron high brilliance X-ray beam, which is produced using focusing optics such as Kirkpatrick-Baez (KB) mirrors at synchrotron facilities. Data are recorded as raster scans of an array of Laue diffraction patterns by a large area charge couple device X-ray detector, and then subsequently analyzed to give grain orientation, stress/ strain distributions, and dislocation densities. Wu, et al. [46] and Chen, et al. [47] have used X-ray microdiffraction to study the microstructure and plastic deformation in microelectronic interconnect materials under electromigration. They also studied the bending in large grains by looking the shifting of the Laue pattern which provides grain rotation information. Microdiffraction provided the insight into the mechanism of plastic deformation by grain rotation [48]. Sarobol, et al. [49–50] have used microdiffraction to study the effect of grain misorientation and elastic anisotropy on tin whisker and hillock formation. They reported that local elastic strain energy density, grain misorientation, and elastic strain gradients influence the formation process.

11.4.1 Microdiffraction Experiment

The X-ray microdiffraction measurements were performed on beamline 12.3.2 at the Advanced Light Source (ALS) in Lawrence Berkeley Laboratory (LBNL). The energy range of the X-ray beam varied from 5 keV to 24 keV. The X-ray beam was focused by Kirkpatrick-Baez (KB) mirrors, and the beam size was approximately 1 μm × 1 μm. To understand the effects of aging, the same region on a joint was scanned for different aging times. A 150 μm × 150 μm square area was analyzed, and Focused Ion Beam milling was used to mark this square area on each joint. The solder joint samples were placed on a high precision stage and scanned through the X-ray beam at its focus spot. Each sample was raster scanned with 1 μm step size and a 1 s exposure time. After initial scanning, the samples were isothermally aged at 125 °C for 1 day and then scanned again. Finally, the samples were aged another 6 days at 125 °C (for a total aging time of 7 days), and scanned a final time. At each scanning step, the Laue diffraction patterns were collected in reflection geometry, using a MAR133 X-ray CCD detector mounted at 90° to the incident beam. The Laue patterns were analyzed and post processed using the in-house XMAS software.

11.4.2 Microdiffraction Results

Figure 11.14 shows the evolution of the subgrain orientations of the solder joint with aging. From these orientation maps, it appears that the subgrains changed significantly for 7 days of aging. By tracking individual subgrains, we found that the subgrains rotated from 0.2644 deg to 0.4602 deg aging, which is believed to be due to shear stress gradients. Grain rotations involve the rearrangement of dislocations that leads to subgrain formation. It is also observed from the orientation map that the subgrain size coarsened with aging. Rotation and formation of subgrains lead to local plastic deformation, which is

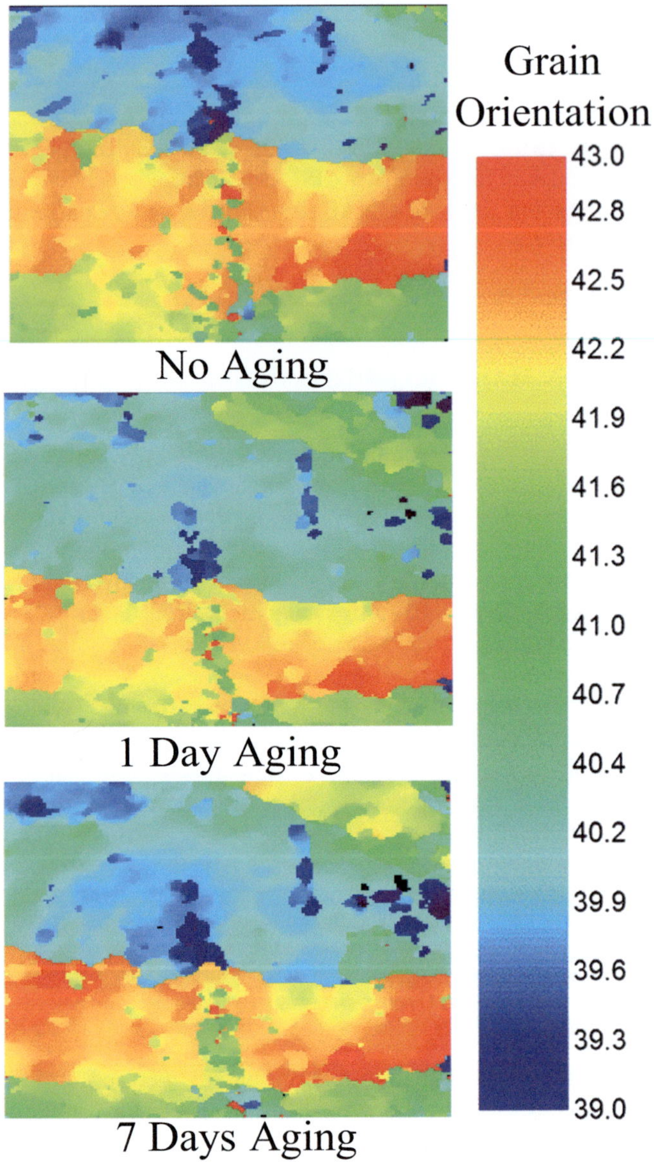

Fig. 11.14 Grain orientation distribution from microdiffraction (SAC305)

diffusion controlled and requires higher temperature. However, deformation mechanisms can be driven by local stresses and stress gradients that cause the creep deformations in a material.

Using micro-diffraction, we were able to calculate the deviatoric stress and strain components, and how they changes with isothermal aging. For example, Figure 11.15 shows a typical result for the deviatoric shear stress components. In all cases, the shear and normal strain components are changing with isothermal aging, and can create stress gradients within the surrounding grains to change the strain energy. This might be the driving force to rotate the grains, cause local changes in the mechanical behavior, and lower the solder joints strength with isothermal aging. Figure 11.16 shows the variation of equivalent strain with isothermal aging. The strain distribution shows a considerable changes with aging. An increasing trend in equivalent strain also indicates the presence of plastic deformation.

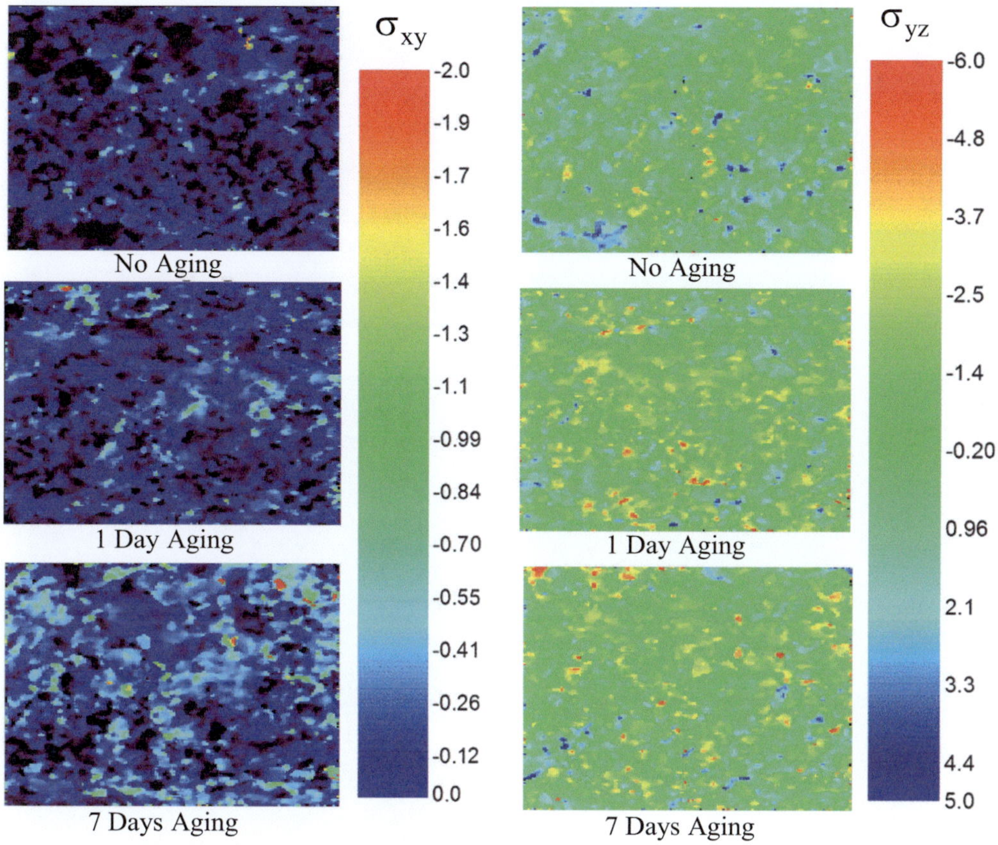

Fig. 11.15 Deviatoric stress distribution from microdiffraction (SAC305)

11.5 Summary and Conclusions

In this investigation, we have extended our previous work on nanoindentation of joints to examine a full test matrix of SAC solder alloys. The effects of silver content on SAC solder aging has been evaluated by testing joints from SACN05 (SAC105, SAC205, SAC305, and SAC405) test boards assembled with the same reflow profile. The observed aging effects in the SACN05 solder joints have been quantified and correlated with the magnitudes observed in tensile testing of miniature bulk specimens performed in prior studies. The results show that the aging induced degradations of the mechanical properties (modulus, hardness) in the SAC joints were of similar order (30-40 %) as those seen previously in the testing of larger "bulk" uniaxial solder specimens. The creep rates of the various tested SACN05 joints were found to increase by 8–50× due to aging. These degradations, while significant, were much less than those observed in larger bulk solder uniaxial tensile specimens with several hundred grains, where the increases ranged from 200× to 10,000× for the various SACN05 alloys. Additional testing has been performed on very small tensile specimens with approximately ten grains, and the aging-induced creep rate degradations found in these specimens were on the same order of magnitude as those observed in the single grain joints. All of the aging effects observed in the SACN05 joints were found to be exacerbated as the silver content in the alloy was reduced.

The enhanced X-ray microdiffraction beamline at the Advanced Light Source (Synchrotron) at the Lawrence Berkeley National Laboratory was also employed to characterize solder joints after various aging exposures (0, 1, and 7 days of aging at $T = 125$ °C). For each case, microdiffraction was used to examine grain growth, grain rotation, sub-grain formation, and residual strain and stress evolution as a function of the aging exposure.

Fig. 11.16 Equivalent strain distribution from microdiffraction (SAC305)

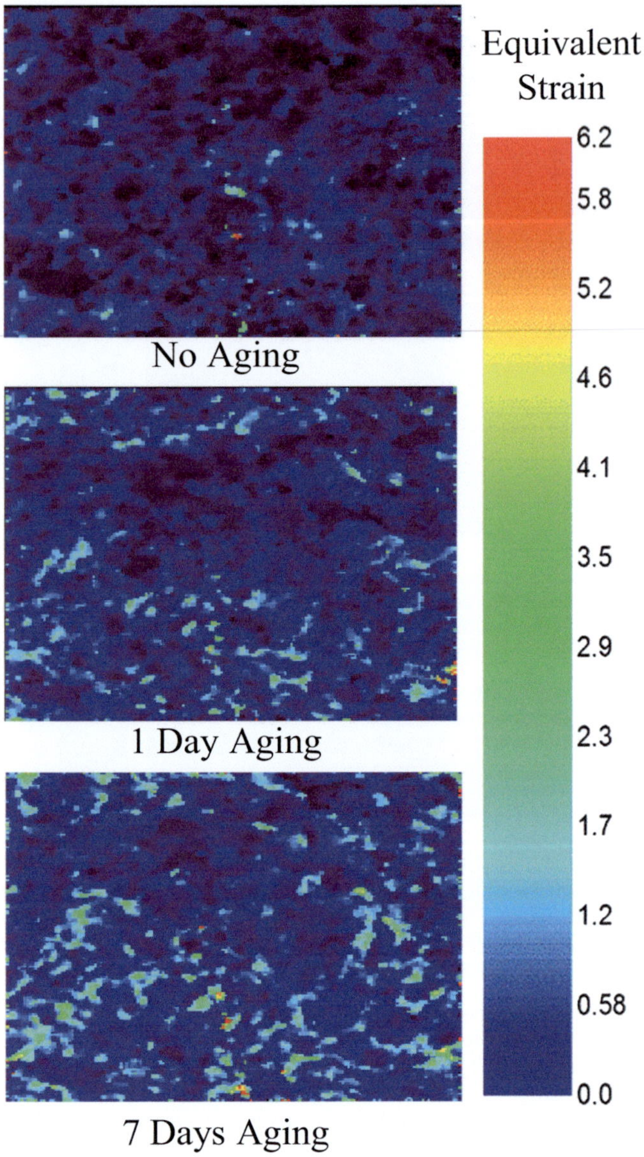

Acknowledgments This work was supported by Center for Advanced Vehicle Electronics and Extreme Environment ($CAVE^3$). We thank the members of iNEMI for providing the joint samples tested in this work. We also gratefully acknowledge and thank Dr. Nobumichi Tamura and Dr. Martin Kunz for assistance with the microdiffraction measurements and data extraction, and for valuable discussions. We acknowledge a grant from the DOE for access to beamline 12.3.2 of the Advanced Light Source (ALS), which is supported by the Office of Basic Energy Science, Material Science Division, of the U.S. Department of Energy under Contact DE-AC02-05Ch11231 at Lawrence Berkeley Laboratory.

References

1. Ma, H., Suhling, J.C.: A review of mechanical properties of lead-free solders for electronic packaging. J. Mater. Sci. **44**, 1141–1158 (2009)
2. Ma, H., Suhling, J.C., Lall, P., Bozack, M.J.: Reliability of the aging lead-free solder joint, pp. 849–864. Proceeding of the 56th Electronic Components and Technology Conference, San Diego, CA (2006)
3. Ma, H., Suhling, J.C., Zhang, Y., Lall, P., Bozack, M.J.: The influence of elevated temperature aging on reliability of lead free solder joints. In: Proceedings of the 57th IEEE Electronic Components and Technology Conference, pp. 653–668, Reno, NV. (May 29-June 1, 2007)
4. Zhang, Y., Cai, Z., Suhling, J.C., Lall, P., Bozack, M.J.: The effects of aging temperature on SAC solder joint material behavior and reliability. In: Proceedings of the 58th IEEE Electronic Components and Technology Conference, pp. 99–112, Orlando, FL. (May 27–30, 2008)

5. Zhang, Y., Cai, Z., Suhling, J.C., Lall, P., Bozack, M.J.: The effects of SAC alloy composition on aging resistance and reliability. In: Proceedings of the 59th IEEE Electronic Components and Technology Conference, pp. 370–389, San Diego, CA. (May 27–29, 2009)
6. Cai, Z., Zhang, Y., Suhling, J.C., Lall, P., Johnson, R.W., Bozack, M.J.: Reduction of lead free solder aging effects using doped SAC alloys. In: Proceedings of the 60th Electronic Components and Technology Conference, pp. 1493–1511. (2010)
7. Mustafa M., Cai Z., Suhling J.C., Lall P.: The effects of aging on the cyclic stress-strain behavior and hysteresis loop evolution of lead free solders. In: Proceedings of the 61st Electronic Components and Technology Conference, pp. 927–939. (2011)
8. Mustafa, M., Cai, Z., Roberts, J.R., Suhling, J.C., Lall, P.: Evolution of the tension/compression and shear cyclic stress-strain behavior of lead-free solder subjected to isothermal aging. In: Proceedings of ITherm 2012, pp. 765–780, San Diego, CA. (May 30–June 1, 2012)
9. Zhang, J., Hai, Z., Thirugnanasambandam, S., Evans, J.L., Bozack, M.J., Sesek, R., Zhang, Y., Suhling, J.C.: Correlation of aging effects on creep rate and reliability in lead free solder joints. SMTA J. **25**(3), 19–28 (2012)
10. Zhang, J., Hai, Z., Thirugnanasambandam, S., Evans, J.L., Bozack, M.J., Zhang, Y., Suhling, J.C.: Thermal aging effects on thermal cycling reliability of lead-free fine pitch packages. IEEE Trans. Component Pack. Technol. **3**(8), 1348–1357 (2013)
11. Motalab, M., Cai, Z., Suhling, J.C., Zhang, J., Evans, J.L., Bozack, M.J., Lall, P.: Improved predictions of lead free solder joint reliability that include aging effects. In: Proceedings of 62nd Electronic Components and Technology Conference, pp. 513–531, San Diego, CA. (May 30–June 1, 2012)
12. Motalab, M., Cai, Z., Suhling, J.C., Zhang, J., Evans, J.L., Bozack, M.J., Lall, P.: Correlation of reliability models including aging effects with thermal cycling reliability data. In: Proceedings of the 63rd IEEE Electronic Components and Technology Conference, pp. 986–1004, Las Vegas, NV. (May 28–31, 2013)
13. Hasnine, M., Mustafa, M., Suhling, J.C., Prorok, B.C., Bozack, M.J., Lall, P.: Characterization of aging effects in lead free solder joints using nanoindentation. In: Proceedings of the 63rd IEEE Electronic Components and Technology Conference, pp. 166–178, Las Vegas, NV. (May 28–31, 2013)
14. Hasnine, M, Suhling, J., Prorok, B.C., Bozack, M.J., Lall, P.: Exploration of aging induced evolution of solder joints using nanoindentation and microdiffraction. In: Proceedings of the 64th Electronic Components and Technology Conference, Orlando, FL, pp. 379–394. (2014)
15. Pang, J.H.L., Low, T.H., Xiong, B.S., Xu, L., Neo, C.C.: Thermal cycling aging effects on Sn–Ag–Cu solder joint microstructure, IMC and strength. Thin Solid Films **462–463**, 370–375 (2004)
16. Darveaux, R.: Shear deformation of lead free solder joints. In: Proceedings of the 55th IEEE Electronic Components and Technology Conference, pp. 882–893 (2005)
17. Wiese, S., Wolter, K.J.: Creep of thermally aged Sn-Ag-Cu solder joints. Microelectron. Reliab. **47**, 223–232 (2007)
18. Dutta, I., Pan, D., Marks, R.A., Jadhav, S.G.: Effect of thermo-mechanically induced microstructural coarsening on the evolution of creep response of SnAg-based microelectronic solders. Mater. Sci. Eng. A **410–411**, 48–52 (2005)
19. Fischer-Cripps, A.C.: Nanoindentation, 3rd edn. Springer, Berlin (2011)
20. Rhee, H., Lucas, J.P., Subramanian, K.N.: Micromechanical characterization of thermo-mechanically fatigued lead-free solder joints. J. Mater. Sci.: Mater. Electron. **13**, 477–484 (2002)
21. Lucas, J.P., Rhee, H., Guo, F., Subramanian, K.N.: Mechanical properties of intermetallic compounds associated with Pb-free solder joints using nanoindentation. J. Electron. Mater. **32**(12), 1375–1383 (2003)
22. Chromik, R.R., Vinci, R.P., Allen, S.L., Notis, M.R.: Measuring the mechanical properties of lead free solder and Sn-based intermetallics by nanoindentation. J. Metals **55**, 66–69 (2003)
23. Deng, X., Chawla, N., Chawla, K.K., Koopman, M.: Deformation behavior of (Cu, Ag)-Sn intemetallics by nanoindentation. Acta Mater. **52**, 4291–4303 (2004)
24. Deng, X., Chawla, N., Chawla, K.K., Koopman, M.: Young modulus of (Cu, Ag)-Sn intemetallics by nanoindentation. Mater. Sci. Eng. A **364**, 240–243 (2004)
25. Gao, F., Taekmoto, T.: Mechanical properties evolution of Sn-3.5Ag based lead free solders by nanoindentation. Mater. Lett. **60**, 2315–2318 (2006)
26. Gao, F., Nishikawa, H., Takemoto, T.: Nanoscale mechanical response of Sn-Ag based lead free solders. In: Proceedings of the IEEE Electronic Components and Technology Conference, pp. 206–210. (2007)
27. Sun, Y., Liang, J., Xu, J.H., Wang, G., Li, X.: Nanoindentation for measuring individual phase mechanical properties of lead free solder alloy. J. Mater. Sci.: Mater. Electron. **19**, 514–521 (2008)
28. Liu, Y.C., Teo, J.W.R., Tung, S.K., Lam, K.H.: High temperature creep and hardness of Eutectic 80Au/20Sn solder. J. Alloys Compd. **448**, 340–343 (2008)
29. Gao, F., Nishikawa, H., Takemoto, T., Qu, J.: Mechanical properties versus temperature relation of individual phase in Sn-3.0Ag–0.5Cu lead free solder alloy. Microelectron Reliab **49**, 296–302 (2009)
30. Han, Y.D., Jing, H.Y., Nai, S.M.L., Xu, L.Y., Tan, C.M., Wei, J.: Temperature dependence of creep and hardness of Sn-Ag-Cu lead free solders. J. Electron. Mater. **39**(2), 223–229 (2010)
31. Han, Y.D., Jing, H.Y., Nai, S.M.L., Xu, L.Y., Tan, C.M., Wei, J.: Indentation size effect on the creep behavior of SnAgCu solder. Int. J. Modern Phys. B **24**(1-2), 267–275 (2010)
32. Xu, L., Pang, J.H.L.: Nanoindentation on SnAgCu solder joints and analysis. J. Electron. Mater. **35**(12), 2107–2115 (2006)
33. Xu, L., Pang, J.H.L.: Nanoindentation characterization of Ni-Cu-Sn IMC layer subjected to isothermal aging. Thin Solid Films **504**, 362–366 (2006)
34. Song, J.M., Huang, B.R., Liu, C.Y., Lai, Y.S., Chiu, Y.T., Huang, T.W.: Nanomechanical response of intermetallic phase at the solder joint interface-crystal orientation and metallurgical effects. Mater. Sci. Eng. A **534**, 53–59 (2012)
35. Venkatadri, V., Yin, L., Xing, Y., Cotts, E., Srihari, K., Borgesen, P.: Accelerating the effects of aging on reliability of lead free solder joints in a quantitative fashion. In: Proceedings of the IEEE Electronic Components and Technology Conference, pp.398–405. (2009)
36. Henshall, G.: iNEMI Lead-Free Alloy Alternatives Project Report: Thermal fatigue experiments and alloy test requirements. In: Proceedings of the SMTAI, pp. 317–324. (2009)

37. Pharr, G.M., Oliver, W.C., Brotzen, F.R.: On the generality of the relationship among contact stiffness, contact area, and elastic modulus during indentation. J. Mater. Res. **7**(3), 613–617 (1992)
38. Oliver, W.C., Pharr, G.M.: An improved technique for determining hardness and elastic modulus using load and displacement sensing indentation experiments. J. Mater. Res. **7**(6), 1564–1583 (1992)
39. Hay, J., Agee, P., Herbert, E.: Continuous stiffness measurement during instrumented indentation testing. Experimental Techniques **34**(3), 86–94 (2010)
40. Lee, B.Z., Lee, D.N.: Spontaneous growth mechanism of Tin Whiskers. Acta Mater. **46**(10), 3701–3714 (1998)
41. Bieler, T.R., Jiang, H., Lehman, L.P., Kirkpatrick, T., Cotts, E.J., Nandagopal, B.: Influence of Sn grain size and orientation in the thermomechanical response and reliability of Pb-free solder joints. IEEE Trans. Component Packag. Technol. **31**(2), 370–381 (2008)
42. Telang, T.U., Bieler, T.R.: The orientation imaging microscopy of lead free Sn-Ag solder joints. J. Metal **57**, 44–49 (2005)
43. Yang, S., Tian, Y., Wang, C.: Investigation on Sn grain number and crystal orientation in the Sn-Ag-Cu/Cu solder joints of different sizes. J. Mater. Sci.: Mater. Electron. **21**, 1174–1180 (2010)
44. Tabor, D.: Hardness of metals. Oxford University Press, Oxford (1951)
45. Kunz, M., Tamura, N., Chen, K., Macdowell, A.A., Ustundag, E.: A dedicated Superband X-Ray microdiffraction beamline for materials, Geo-, and environmental sciences at the advanced light source. Rev. Sci Instrum. **80** (2009)
46. Wu, A.T., Tamura, N., Lloyd, J.R., Kao, C.R., Tu, K.N.: Synchrotron X-ray microdiffraction analysis on microstructure evolution in Sn under electromigration. Mater. Res. Soc. Symp. Proc. **863**, (2005)
47. Chen, K., Tamura, N., Valek, B.C., Tu, K.N.: Plastic deformation in Al(Cu) interconnects stressed by electromigration and studied by synchrotron polychromatic X-ray icrodiffraction. J. Appl. Phys. **104**, 013513 (2008)
48. Kunz, M., Chen, K., Tamura, N., Wenk, H.R.: Evidence of residual elastic strain in deformed natural quartz. Am. Mineralogis. **94**, 1059–1062 (2009)
49. Sarobol, P., Koppes, J.P., Chen, W.H., Su, P., Blendell, J.E., Handwerker, C.A.: Recrystallization as a nucleation mechanism for Whiskers and Hillocks on thermally cycled Sn-alloy solder films. Mater. Lett. **99**, 76–80 (2013)
50. Sarobol, P., Chen, W.H.: Effects of local grain misorientation and β-tin elastic anisotropy on Whisker and Hillock formation. J. Mater. Res. **28** (5), 747–756 (2013)

Chapter 12
High Rate Experimental Test Method for Harsh Environment Stretchable Electronics

Ryan Lowe, Steve Dooley, Dan Berrigan, and Jason Foley

Abstract Stretchable electronics that can bend, conform, and survive high mechanical strains have the potential to become a paradigm changing innovation for harsh environment electronics subjected to extreme drop/shock, vibration, and temperature loads. Traditional electronic packaging best practices do not necessarily apply to stretchable electronics. Models of many stretchable electronics have yet to be validated experimentally due to the variety of complicated interfaces/materials. Furthermore test standards commonly used with FR-4 based electrical systems, like JEDEC/JESD22B-111 (Board Level Drop Test for Handheld Electrical Products) are not well suited to evaluate stretchable electronics. This paper motivates the need for new engineering tools and discusses the development of a new experimental test method designed specifically for the development of flexible harsh environment electronics. The new method fills a capability gap between constant rate material characterization test methods used to evaluate properties of constituent materials, and full system level tests that make isolating a single components failure mechanism difficult. The presented method is an attempt to create an inexpensive, quick, and convenient test for subjecting component level test articles to strain in excess of 100 % at non-constant strain rates on the order of 300/s or peak accelerations in excess of 60,000g's. Advantages and limitations of the technique are discussed.

Keywords Electronics • Harsh environment • Flexible • Stretchable • High-g

12.1 Introduction

Stretchable electronics which vary fundamentally in constituent materials from flex circuits and flexible PCB's are primarily research grade assemblies that have few widely commercialized realizations at the time of this writing [1]. Stretchable electronics are expected to bend, fold, and conform like other flexible circuits, but also tolerate high levels of mechanical strain. Compared to standard FR-4/370HR glass reinforced epoxy circuit boards ($E \approx 1525$ GPa [2]) commonly used in consumer electronics applications, stretchable electronics substrates have a modulus of elasticity orders of magnitude lower ($E \approx 550$ MPa [3]). The difference in material properties and intended use environments motivates this work and the creation of a new test methods. Small strain levels less than 0. 1 % in a traditional circuit board can cause component level failures at solder joints [4]. In contrast flexible electronics are designed from inception to survive very high ($\approx$ 100 %) strain levels [5]. For example flexible electronics that are applied as a patch to human skin for performance monitoring stretch and wrinkle along with the skin during activity. The use of stretchable electronics for harsh shock and vibration environments is still a young research area, and new test methods are required to fully understand its potential. Classical constant rate material testing methods can be used to understand the materials used in stretchable electronics under conditions considered harsh. But, currently there is no high rate test capability for evaluating stretchable electronic assemblies comprised of a substrate and a few components for the purpose of understanding assembly level survivability. This paper describes an effort to develop an engineering test to study and characterize component level reliability of stretchable electronics.

R. Lowe (✉)
Applied Research Associates, Littleton, CO, USA
e-mail: rdlowe@ara.com

S. Dooley
Air Force Research Laboratory Sensors Directorate, Wright-Patterson Air Force Base, Dayton, OH, USA

D. Berrigan
Air Force Research Laboratory Materials Directorate, Wright-Patterson Air Force Base, Dayton, OH, USA

J. Foley
Air Force Research Laboratory, Eglin Air Force Base, Okaloosa, FL, USA

B.C. Prorok, L. Starman (eds.), *MEMS and Nanotechnology, Volume 5*,
Conference Proceedings of the Society for Experimental Mechanics Series, DOI 10.1007/978-3-319-22458-9_12

The test methodology is loosely derived from the traditional JEDEC standard test for consumer electronics which utilizes a drop tower for testing [6]. Like the JEDEC test the new methodology allows for component level failures to be localized and studied in a controlled laboratory test. Furthermore the test has no direct correlation to a specific usage environment, rather it is a generalization of the types of forces that can fail electronics. While the test method may look like a classical constant rate tensile test, the strain rate is not constant, and is closer in frequency content to actual harsh shock/impact loads that could be experienced in a fielded system. The operating principle of the test is derived from a split-Hopkinson (Kolsky) bar test with the use of a pneumatic air gun, but the analysis of the test does not consider stress waves directly. In the following sections the design and operation of the test machine is described along with specifics of the test method. Finally experimental results are presented from initial test specimens to highlight advantages and limitations of the technique. Due to the nature of the violent impact needed to create high rate loads this test has been dubbed a pop shock test.

12.2 Theory of Operation

The operating principle of the pop shock test is to fire a steel projectile into a transmission rod that then loads a specimen. The test is generically defined and could incorporate elements of tension, compression, or shear depending on test design. The pneumatic gun creates the energy for the test. The test area is designed to be as customizable as possible and is individually tailored to meet the needs of each test. Figure 12.1 shows an overall view of the pop shock tester.

The study of internal ballistics has a long history [7], but it was unclear from the few papers [8, 9] that described the design of a low pressure ($<$ 100 psi) air gun how appropriate the simple model shown in Fig. 12.2 would capture the physics of a gun when using cheap loose tolerance components. It was also unclear how sensitive the design would be to the speed at which the solenoid valve opened. Lastly clearance is needed between the striker and the barrel to ensure free travel of the projectile, which could allow air to escape around the striker resulting in lower velocities. All together there were many practical design considerations that could possibly reduce the total energy developed during a test.

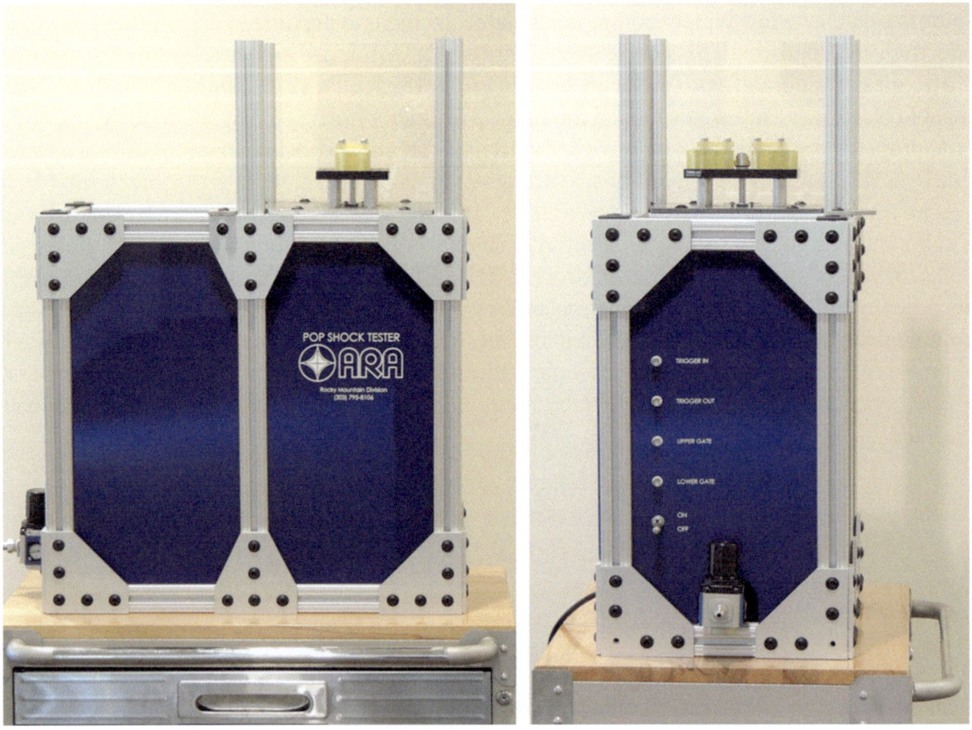

Fig. 12.1 Test machine shown from front and side. The pneumatic gun is enclosed behind the panels, while the test area is visible behind clear protective panels at the top

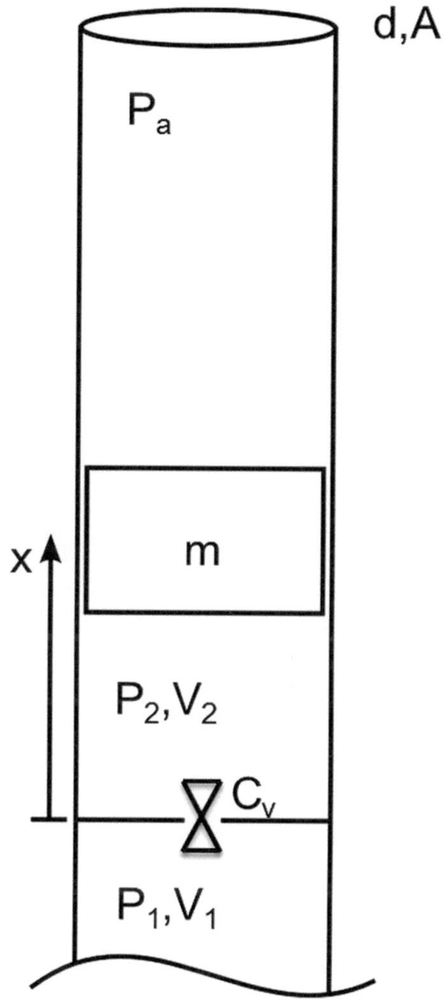

Fig. 12.2 Simple first principles model of internal ballistics

A simple first principle model for the pneumatic gun is shown in Fig. 12.2. The absolute pressure in the reservoir is denoted as P_1, with a volume of V_1. When the gun is fired a solenoid valve with a flow coefficient, C_v, allows air into the volume, V_2, at pressure P_2 behind the striker of mass m and cross sectional area A. The striker is assumed to perfectly fit inside the barrel of diameter d. Friction at the striker/barrel interface is denoted as f. The atmosphere is at a pressure of P_a.

The equation of motion of the striker can be modeled as:

$$\frac{d^2x}{dt^2} = 1/m(P_2A - P_aA - f - mg) \tag{12.1}$$

Where x is the coordinate representing the strikers displacement along the barrel, and g is acceleration due to gravity. The ideal gas law is used to model the gas expansion in the reservoir and barrel. Applying the ideal gas law to the reservoir results in Eq. (12.2).

$$P_1(t)V_1 = N_1k_bT \tag{12.2}$$

Where N_1 represents the number of molecules in the reservoir, k_b is Boltzman's constant, and T is the temperature of the gas (assumed adiabatic at room temperature). Applying the ideal gas law to the volume behind the striker results in Eq. (12.3).

$$P_2(t)(Ax(t)) = N_2k_bT \tag{12.3}$$

Where N_2 represents the number of molecules in the reservoir. The number of molecules in the reservoir and barrel respectively is a function of the molecular flow rate, Q between the two volumes. Assuming conservation of mass the molecular flow rate, Q, into each volume is,

$$\frac{dN_1}{dt} = -Q \tag{12.4}$$

$$\frac{dN_2}{dt} = Q \tag{12.5}$$

The flow of air through a valve has two regimes that are a function of the pressure drop [10]. At low pressure drops where $P_2 < P_1/2$ equation (12.6) governs the flow through the valve and the outlet pressure restricts flow. At high pressure drops the flow of gas is a linear function of the inlet pressure and is governed by Eq. (12.7). This condition is commonly denoted as choked flow.

$$Q = C_c C_v P_1 \left(1 - \frac{2\Delta P}{3P_1}\right)\sqrt{\frac{\Delta P}{P_1 G_g T}} \tag{12.6}$$

$$Q = (0.471) C_c C_v P_1 \sqrt{\frac{1}{G_g T}} \tag{12.7}$$

where C_c is a constant value of 6950, and G_g is the specific gravity of air. Equations (12.6) and (12.7) are approximations derived from ISA S75.01. By taking time derivatives of Eqs. (12.2) and (12.3), the governing differential equations for the state vector $(x, \dot{x}, P_1, P_2)$ can be written as:

$$\begin{Bmatrix} \dot{x} \\ \ddot{x} \\ \dot{P}_1 \\ \dot{P}_2 \end{Bmatrix} = \begin{Bmatrix} \dot{x} \\ 1/m(P_2A - P_aA - f - g) \\ -Q\dfrac{k_bT}{V_1} \\ \dfrac{Qk_bT - P_2A\dot{x}}{V_2} \end{Bmatrix} \tag{12.8}$$

Using initial conditions for $x, \dot{x}, P_1,$ and P_2 the system of equations can be solved numerically. Within a rough approximation the model described by Eq. (12.8) provide a method for estimating the velocity of a striker fired from a pneumatic gun. Unmodeled losses that affect the theoretical vs actual speeds developed by the striker during testing are presented in Fig. 12.7 of the experimental results section.

The test area can be used to create a variety of different test conditions. This paper will focus on a tension test that stretches test articles. During testing the pneumatic gun fires the projectile vertically into a transmission bar. The impact causes the transmission bar to move up vertically stretching the test article. Figure 12.3 shows the progression of the striker up the barrel and the resulting impact with the transmission bar. For tension testing Fig. 12.4 shows the fixture that is attached to the test area. As the transmission bar moves vertically the lower edge of the test specimen stays stationary resulting in a tensile loading. Other fixtures for shear and more complicated loadings are also possible. For example Fig. 12.1 shows the pop shock tester with the flexure fixture installed.

12.3 Test Method

For the testing presented in this paper a 10.2 cm (4 in.) steel striker with a diameter of 1.9 cm (0.75 in.) was used as the projectile in the pneumatic gun. Felt pads (extra firm F1 pure wool, McMaster-Carr) of 0.635 cm (0.25 in.) where used as shock programmers between the striker and transmission bar. Air pressures ranging between 10 and 100 psi were used during testing.

Optical gates (photo transistor emitter/receiver pair, Sparkfun Electronics) on the pop shock tester are aligned over the vent at the top of the barrel. After the impact of the striker with the bottom of the transmission bar the transmission bar rises and blocks the optical gates in succession.

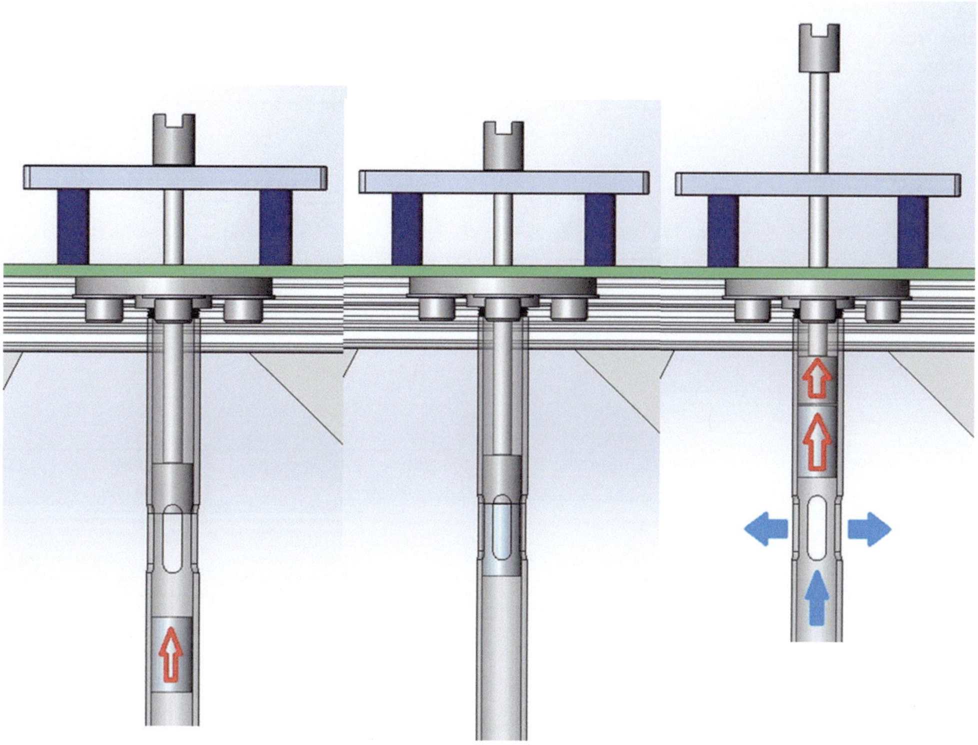

Fig. 12.3 Transmission of energy from the pneumatic gun into the test area through the transmission bar. The transmission bar attached to a test fixture to power the test

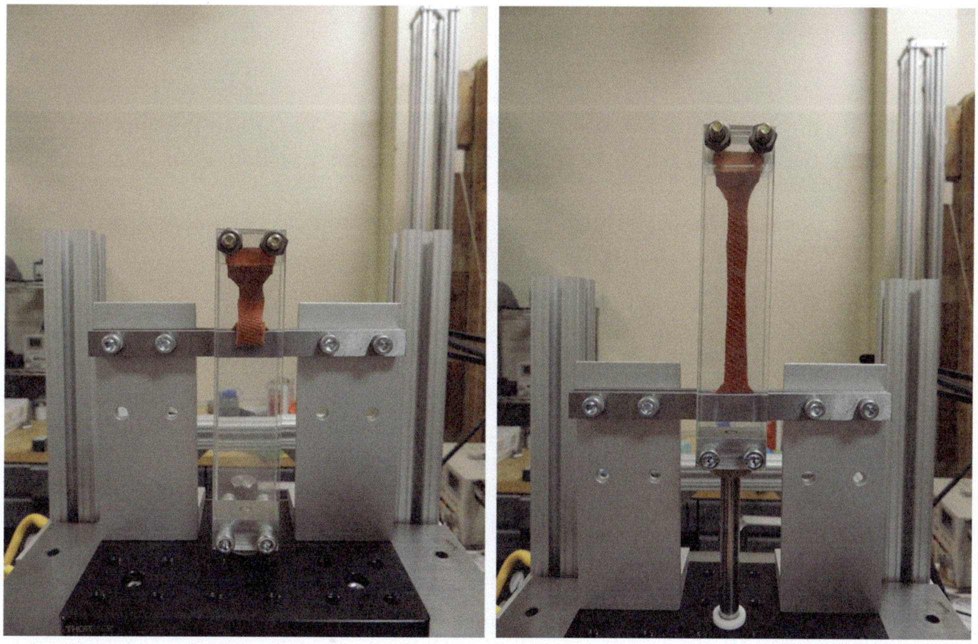

Fig. 12.4 Tension test fixture before and after test

The voltage signals from the optical gates is nominally 1 V when the gate is open, and 5 V when the gate is closed (obstructed). Their are two gates on the pop shock tester. The lower gate is closest to the bottom of the machine and closes as the transmission bar passes vertically past it. The upper gate is always the second gate to close. An arbitrary threshold value of 4.5 V is used to define when a gate is closed. The optical gate data was post processed using the following steps:

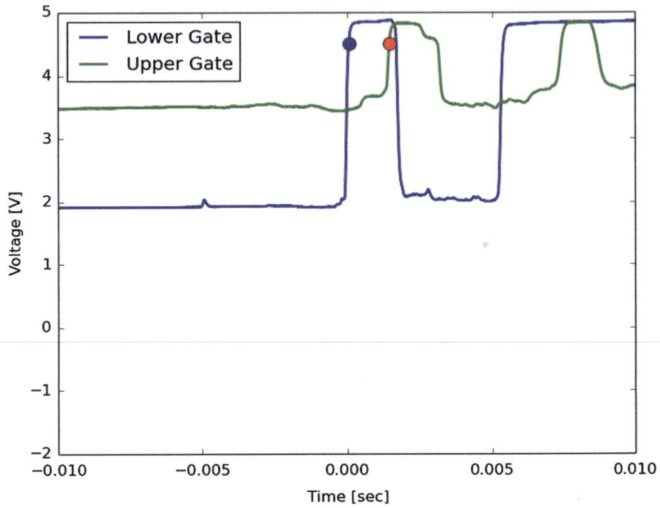

Fig. 12.5 Nominal data showing voltage output from optical gates. The voltage threshold used to indicate a closed gate is shown with round markers

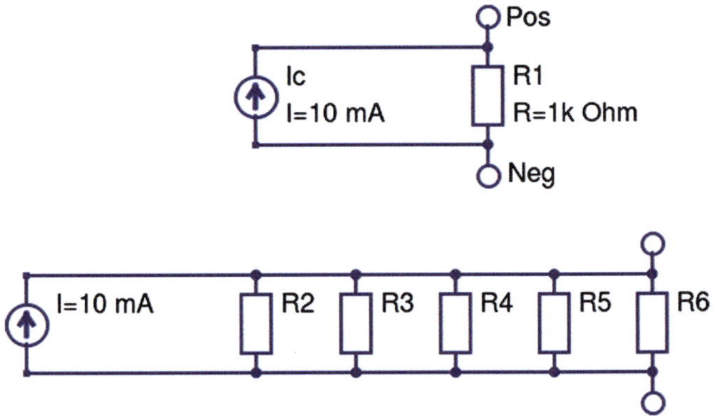

Fig. 12.6 Schematic of daisy chain circuits for a single component (*top*), and multiple components (*bottom*)

1. For the lower gate, identify the first data point above the threshold of 4.5 V. Denote this time t_{lower}
2. For the upper gate, identify the first data point above the threshold of 4.5 V. Denote this time t_{upper}
3. Calculate $\Delta t = t_{upper} - t_{lower}$
4. The distance between the gates is closely controlled with a laser cut fixture and denoted as $\Delta d = 2$ cm
5. Velocity of the transmission bar is calculated as $V_{gate} = \frac{\Delta d}{\Delta t}$

An observed voltage time history from the optical gates output is shown in Fig. 12.5. The upper gate is last to close. The blue and red dots in Fig. 12.5 indicate the first point in time when the signal reached 4.5 V for each signal. The time difference between the points is $\Delta t = 1.39$ ms. Using consistent units, the velocity of the transmission bar can be calculated as $v = (0.02\text{ m})/(0.00139\text{ s}) = 14.3$ m/s. To verify the calculation of velocity a shock accelerometer (Endevco 7280AM4-60k) was mounted to the top of the transmission bar. For peak accelerations less than 50,000g the error between the integrated acceleration and velocity calculated with the optical gates was less than 5 %. For accelerations above 60,000g the velocity error starts to increase above 5 %. We believe this error is related to the excitation of the resonant frequency of the accelerometer at 100 kHz, but further study is required to fully understand the discrepancy in predicted velocities at the upper limit of the machine.

One of the motivations for developing this test method was to enable diagnostic methods for failures in flexible electronics. For simple resistive components daisy chain circuits can be monitored in-situ with high speed data acquisition for intermittent open events. Figure 12.6 shows a typical daisy chain circuit used for in-situ monitoring. The voltage between the positive (pos) and negative (neg) ports will change drastically in the event of a component failure that causes a change in resistance. In the event that the number of components is greater than the number of available data acquisition channels parallel circuits can be implemented to accommodate larger numbers of components. The trade-off between daisy chain circuit topologies is a reduced ability to localize failure locations when parallel circuits are implemented. Other types of components such as capacitors are more difficult to monitor in-situ for changes in electrical properties, but the number of possible approaches are increasing [11]. One motivating factor in the design of the tension fixture was the ability to keep one end of the specimen stationary through out the test. Connecting and maintaining electrical connections to test articles during the dynamic test is easier with one end of the test article remaining stationary.

12.4 Experimental Results

With an emphasis on a cheap and quick test capability it was initially not clear how loose tolerances in the barrel (water pipe, McMaster-Carr) would effect the actual striker speed. Figure 12.7 highlights the deviation of actual striker speeds from theoretical for a range of pressure set points. The resulting average velocity error between theory and actual over the range of pressures was about 30 %.

For the purpose of test planning an operating curve was generated to map the pressure of the pneumatic gun reservoir to the speed of the transmission bar. Figure 12.8 shows the maximum transmission bar velocity as ~ 14 m/s (46 ft/s).

Sample size and transmission bar velocity control the strain and average strain rate possible with the pop shock tester. The minimum sample size, L_i, is 3.18 cm (1.25 in.). At full extension the sample is stretched to a length, L_f, of 10.80 cm (4.25 in.). Therefore the maximum engineering strain that can be achieved is $\epsilon_{max} = (L_f - Li)/Li = (10.8 - 3.18)/3.18 = 2.4 = 240\ \%$. Test articles tend to be longer in practice ($\approx$ 5 cm), so

Fig. 12.7 Actual striker speed compared to the model described in Fig. 12.2

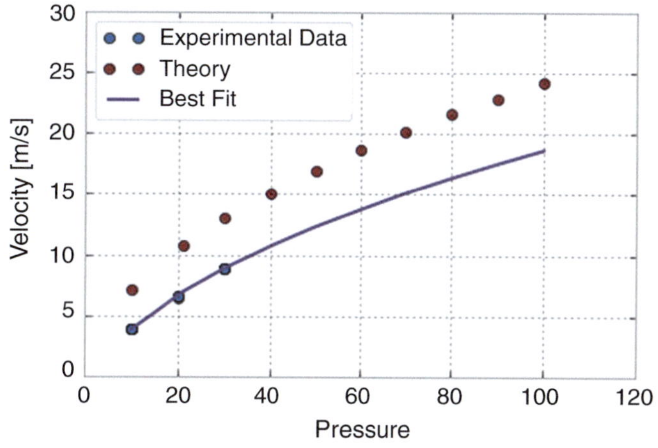

Fig. 12.8 Operating curve mapping pneumatic gun pressure to transmission bar velocity

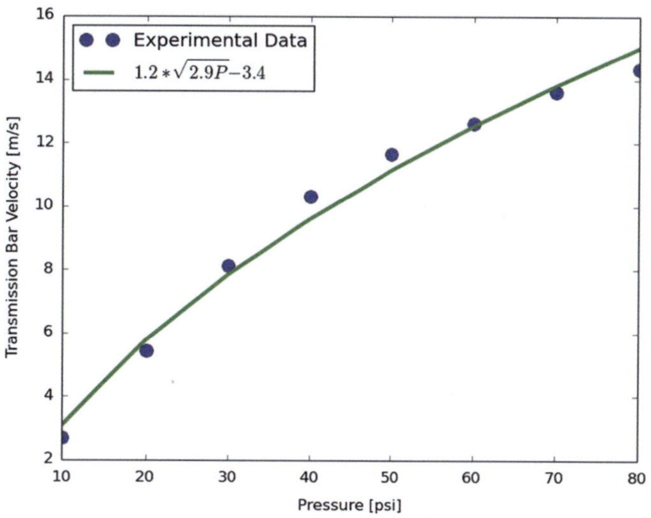

practical strains achieved are usually closer to 100 % strain. The dynamics of stressing test samples generates forces that cause the strain rate to no be constant, but an average engineering strain rate can be approximated as $\dot{\epsilon}_{max} = V_{t-bar}/L_i = (14\ \text{m/s})/(0.0318\ \text{m}) = 440/\text{s}$, where V_{t-bar} is the velocity of the transmission bar. Typical sized samples see an average strain rate of 280/s.

12.5 Advantages and Limitations

The pop shock tester is advantageous because it can provide a benchmark to measure the relative survivability of stretchable electronics from different materials, manufacturing methods, and geometries. It is also capable of large strains over 100 % and intermediate strain rates on the order of 300/s. Assembled from easy to source components with loose tolerance levels the pop shock tester provides a quick, cheap, and easy way to repeatably load stretchable electronics.

A few of the disadvantages of the pop shock tester are the non-uniform strain rate applied to the specimen, and the lack of scale factors that relate a test to an actual usage environment. Additionally there is currently only a single version of the pop shock tester in existence, so comparisons of results between different laboratories is not yet possible.

12.6 Conclusion

This paper presented a test method that was designed to provide data for engineering judgments of stretchable electronics. The method is more basic than system level testing with fully fielded devices, but more application realistic than classical material characterization methods. Furthermore the test provides the ability to test assemblies at relatively high rates. A comparison was made between a simple theoretical model that predicts test velocity, and experimental observations. The paper focused on a test methodology using a tension fixture, but a variety of test configurations are possible to perform. Instrumentation for both the pop shock tester, and in-situ diagnostics for samples under test were discussed.

Acknowledgements This work was supported by Air Force Research Laboratory contract number FA8651-11-D-0046.

References

1. Rogers, J., et al.: Materials and designs for wireless epidermal sensors of hydration and strain. Adv. Funct. Mater. **24**, 3846–3854 (2014)
2. Lall, P., Panchagade, D., Liu, Y., Johnson, W., Suhling, J.: Models for reliability prediction of fine pitch BGAs and CSPs in shock and drop impact. IEEE Trans. Compon. Packag. Technol. **29**(3), 464–474 (2006)
3. Larmagnac, A., Eggenberger, S., Janossy, H., Voros, J.: Stretchable electronics based on Ag-PDMS composites, Scientific Reports No 4 (2014)
4. Lall, P., Panchagade, D., Iyengar, D., Snadeep, S., Suhling, J., Schrier, H.: High speed digital image correlation for transient-shock reliability of electronics. In: Electronic Components and Technology Conference (2007)
5. Rogers, J., et al.: Fractal design concepts for stretchable electronics. Nat. Commun. (2013).
6. JEDEC Solid State Technology Association: Board-level drop test method of components for handheld electronic products, No JESD22-B111 (2003)
7. Noble, A.: Artillery and Explosives. EP Dutton, New York (1906)
8. Rohrbach, Z., Buresh, T., Madsen, M.: Modeling the exit velocity of compressed air cannon. Am. J. Phys. **80**(1), 24–26 (2012)
9. Tassa, Y., Wu, T., Movellan, J., Todorov, E.: Modeling and identification of pneumatic actuators. In: IEEE International Conference on Mechatronics and Automation (ICMA) (2013)
10. Swagelok valve sizing technical bulletin, Swagelok Inc., MS-06-84-E (2007)
11. Kaplan, S., Greendyke, R., Lowe, R., Foley, J.: Measurement of electromagnetic response of capacitors in dynamic loading conditions. In: Society of Experimental Mechanics Conference (2015)

Chapter 13
Measurement of the Electromechanical Response of Capacitors in Dynamic Loading Conditions

Sergey M. Kaplan, Robert B. Greendyke, Ryan D. Lowe, and Jason R. Foley

Abstract Capacitors are common discrete components in electronics, including systems intended for operation in harsh environments. However, capacitance response due to the operating environment has not been well characterized in the literature. This work describes a technique to simultaneously evaluate the electrical and mechanical responses of capacitors subjected to dynamic mechanical loading by measuring capacitance through the calculation of the RC time constant. Printed circuit boards with surface mounted capacitors were subjected to moderate drop shocks using an accelerated drop tower. The capacitance was estimated at a frequency of 100 kHz during and after the impact, by applying a least squares fit to capacitor charge/discharge voltage measurements. Capacitance fluctuations post impact coincided with the board strain oscillations and were subsequently observed to decrease abruptly, which correlated with high speed imaging showing the onset of capacitor failure. The amount of change in capacitance was indicative of the total or partial flexural failure modes of floating electrode capacitors accounting for more than 90 % of breakdowns. The 6 % standard deviation accuracy of the technique did not allow for detection of piezoelectric response, suggesting that even at peak accelerations on the order of 24,000 g the mechanical mode of failure still dominates.

Keywords: High-g • Strain • Real-time • RC constant • Capacitance

13.1 Introduction

A vast majority of commercial applications utilize electronics packages under relatively benign mechanical acceleration environments, and thus the standards for testing electronics exposed to drop impact specified in Joint Electron Device Engineering Council (JEDEC) standards are limited to 2900 g [1]. Therefore, electronics packages and subassemblies covered by the JEDEC testing specifications have been well studied, and have been found to be relatively predictable and reliable under these benign conditions [2–5].

However, surface mounted electrical components are becoming increasingly prolific inside electronics packages. Particularly, in the aerospace applications that traditionally were serviced by mechanical systems and where potentially huge cost savings can be realized with the implementation of electronics packages, the industry is incentivized to increase its dependence on electrical systems. However, the growing use of these electronics packages in aerospace applications corresponds to an ever increasing range of technology exploitation environments. Stress profiles with peak accelerations of 10,000 g and higher, which is at least an order of magnitude greater than specified by JEDEC, are often encountered by crash survivable electronics exposed to extreme acceleration profiles, and space applications, where high relative speeds correspond to high kinetic energy. Electronics used in these moderately high-g impact applications often use surface mounted capacitors as part of their packages, but the failure modes and mechanisms of capacitors exposed to such environments have not been very well understood.

S.M. Kaplan (✉)
Department of Aeronautical Engineering, Air Force Institute of Technology, Wright-Patterson AFB, Dayton, OH 45431, USA
e-mail: sergey.kaplan@us.af.mil

R.B. Greendyke
Department of Aeronautical Engineering, Air Force Institute of Technology, Wright-Patterson AFB, Dayton, OH, USA

R.D. Lowe
Applied Research Associates, Denver, CO, USA

J.R. Foley
Air Force Research Laboratory, Eglin AFB, Okaloosa County, FL, USA

B.C. Prorok, L. Starman (eds.), *MEMS and Nanotechnology, Volume 5*,
Conference Proceedings of the Society for Experimental Mechanics Series, DOI 10.1007/978-3-319-22458-9_13

In order to save costs by reducing weight and size of packages used in aerospace applications, electronics are being increasingly miniaturized, making surface mounted components more abundant. However, as the size of an electrical component such as a capacitor decreases, its capacitance, C, is also reduced based on the volumetric efficiency factor (A/d), as seen in Equation 1, where A is the normally overlapping area of the electrode plates, d is the distance between electrode plates, n is the number of layers of dielectric material and ε_r is the relative permittivity compared to the permittivity of free space, $\varepsilon_0 = 8.854$ F/m [6].

$$C = \varepsilon_0 \varepsilon_r \frac{A}{d} n \tag{13.1}$$

The only way to increase or maintain the capacitance value of Multi Layer Ceramic Capacitors (MLCC), as the size of component is decreased, is through the use of exotic ferroelectric materials, such as Barium Titanate ($BaTiO_3$) and Lead Zirconium Titanate (PZT), which have high relative permittivity on the order of 1250–10,000, depending on operating temperature [6, 7]. Unfortunately, dielectrics based on ceramic materials such as $BaTiO_3$ that are frequently used in MLCC are also piezoelectric and their performance—specifically stability of permittivity in the direction normal to the electrode plates, which affect capacitance—under high mechanical stress and more extreme acceleration profiles than have been traditionally explored, has not been thoroughly investigated.

Recent studies conducted by the capacitor manufacturing industry suggest that capacitance primarily changes due to flex cracking of the dielectric material [8–10]. Under certain static loading conditions such as a bending test, capacitance reverts back to within 1 % of its original value immediately after the stress is removed. This motivates the need for in-situ monitoring during highly dynamic tests in addition to pre/post-test capacitance measurements [9]. Intermittent nature of electronic components is not a new phenomenon and has recently been closely studied as it affects the reliability of Ball Grid Assembly (BGA) packages through the monitoring of resistive loads in an electrical circuit undergoing a maximum acceleration of 3,700 g [2, 3]. However, changes in capacitance during the shock event have not yet been quantified, particularly with acceleration profiles with peak values greater than those encountered in most commercial applications.

Other studies have been performed to investigate the piezoelectric behavior of raw dielectric materials such as PZT under quasi-static conditions and dynamic mass drop experiments. The voltages and currents that are generated by the piezoelectric materials have been closely related to magnitude and type of mechanical load [11]. Also, stress applied to the block of PZT, where during the dynamic stress event the charge built up on the surface does not have the time to dissipate leads to higher currents than during the quasi-static stress tests [12]. However, exploring electromechanical response of raw PZT material under carefully controlled conditions, while a good starting point in understanding electronics behavior under large mechanical stresses, does not address the gap in understanding of reliability issues of capacitors employing ferroelectric dielectric materials under realistic and dynamic exploitation conditions.

While the flex cracking of the capacitors, which is predominantly driven by the mechanical stress response of the materials has been thoroughly studied, and the electromechanical response of some of the raw ferroelectric materials used as a dielectric inside capacitors has been characterized, the relationship between the mechanical dynamic stress conditions, indicative of moderately high-g acceleration profiles, and capacitance has not been investigated. The current work investigates the electromechanical response of capacitors undergoing drop impacts on the order of 24,000 g to better understand failure modes and mechanisms of electronics exposed to large dynamic mechanical stresses. Surface mounted capacitors are monitored using a newly developed high-frequency RC constant measuring technique in order to investigate any piezoelectric effect through a change in capacitance during the shock event and to identify the time of the mechanical failure of the component. Shocked capacitors are then dismounted from the PCB and static capacitance is measured. Correlation between the board dynamics during the drop event and the mode of failure is presented.

13.2 Test Set-Up

13.2.1 Dynamic Capacitance Measurement Methodology

Given a constant potential V_0, through a resistor of resistance, R in Ohms, in series with a capacitor of capacitance C in Farads, voltage across the capacitor will change at a predictable rate given in Eqs. (13.2 and 13.3) respectively, where $\tau = RC$ is time in seconds represented by the x-intercept of the line of initial slope of the current in the circuit. [13]

$$\text{Charge}: \; V(t) = V_0\left(1e^{t/}\right) \tag{13.2}$$

$$\text{Discharge}: \; V(t) = V_0\left(e^{t/}\right) \tag{13.3}$$

By applying a square voltage wave input to the RC circuit during a mechanical shock event and monitoring voltage across the capacitor over time during each charge/discharge, changes in time constant can be determined and, if resistance is held constant, capacitance can be calculated.

13.2.2 Devices Under Test

Capacitors used during this work were AVX Corporation Class II MLCC based on Barium Titanate ceramic dielectric with X7R temperature profiles (relatively low variance of capacitance with temperature). The floating electrode capacitors, shown in Fig. 13.1, were rated to 3000 VDC and 470 pF in capacitance. The nominal size of devices was 4.5 3.2 1.54 mm (size 1812 in inches) with each component weighing 0.12 g.

An Isola 370HR rectangular 60.5 19.2 1.5 mm circuit board housed up to four capacitors on the top surface of masked copper stations C5, C6, C7 and C8 with up to four connections per station, as seen in Fig. 13.2. Capacitors were surface mounted on the PCB with the Exmore VS-500 Vapor Soldering reflow oven according to AVX surface mounting guide temperature profile using the Sn63Pb37 Kester Manufacturer solder paste applied to the PCB copper pads by the Mann Corp SP5500 Dual Squeegee Stencil Pencil [14].

13.2.3 Data Acquisition Equipment

A 3.048 m (10 ft) MTS drop tower apparatus assisted with a bungee cord and 1/8 in. thickness programmer material to shape the acceleration pulse was used during drop impact experiments. Initial mechanical shock impulse duration was on the order of 100–250 μs with peak acceleration amplitudes ranging between 500 and 24,000 g depending on the drop height.

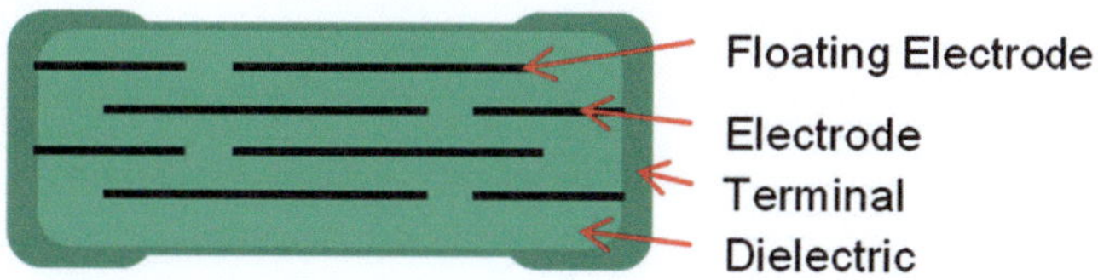

Fig. 13.1 Floating electrode MLCC

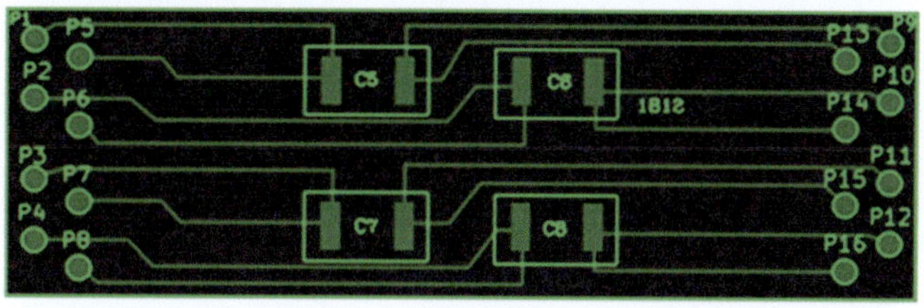

Fig. 13.2 Printed circuit board layout

The charge/discharge square pulse was generated by waveform generators at a frequency of 100 kHz during the drop resulting in approximately 10–25 capacitance measurements throughout the duration of the initial shock impulse. Capacitance was later calculated using least squares fit of an exponential function to the voltage versus time curve.

The master charge/discharge signal was provided by the Standard Research Systems DS345 Synthesized Function Generator using a 5 VDC peak to peak square wave with zero offset routed through the off-board, unstressed, capacitor via the shorted C8 station on the PCB and a 464 Ω resistor. The slave charge/discharge signal was provided by the Agilent 33220A Arbitrary Waveform Generator using the 5 VDC peak to peak square wave through the stressed capacitor on C5 station on the PCB and a 464 Ω resistor. TDSK5104B Digital Phosphor Oscilloscope was used to monitor and record voltage across both stressed (Chap. 3) and unstressed (Chap. 4) capacitors, as well as the original slave (Chap. 1) and master (Chap. 2) square pulses. Voltage across the mechanically stressed capacitor was measured and recorded to observe the direct effects of the mechanical shock on the capacitance, while voltage across unstressed, off-board, capacitor was measured for comparison in order to detect any capacitance changes due to cable whip, and inherent capacitance in the cabling and the board. The slave and master pulse generators were synchronized and square waves were routed through the same 20 ft of commercial grade general purpose 22 AWG 4 conductor plenum shielded cable, with the stressed capacitor on C5 station using+red, −black convention and short on C8 PCB station with off-board capacitor using+green, −white coding. A Vishay foil strain gage was fixed to the bottom surface of the PCB to measure the lengthwise (x-direction) uniaxial strain response of the board during the impact in order to quantify the board dynamics. An Endevco 7270A-60 K SNF30156 accelerometer was mounted directly on the fixture to measure impact acceleration in the vertical direction (z-direction). The accelerometer signal was routed through the Auxiliary oscilloscope channel to trigger recording of capacitance data on all four channels with a 15–25 % delay, once the voltage of 140 mV, equivalent to 280 g, was reached. National Instruments Chassis supplied constant current for the strain gage and constant voltage for the accelerometer with Precision Filters 28000 providing initial signal processing as part of the impact data acquisition system. The wring schematic for the data acquisition system is shown in Fig. 13.3.

Acceleration and strain data were recorded at a sample rate of 25 MS/s (25 MHz) for a total duration of 100 ms, while the voltage across both stressed and unstressed capacitors, as well as the original slave and master signals was collected at a sample rate of 50 MS/s (50 MHz) for a total duration of 10 ms. With the 100 kHz wave pulse frequency, approximately 250 voltage data points were available for the least squares fit to an exponential function in order to determine the RC time

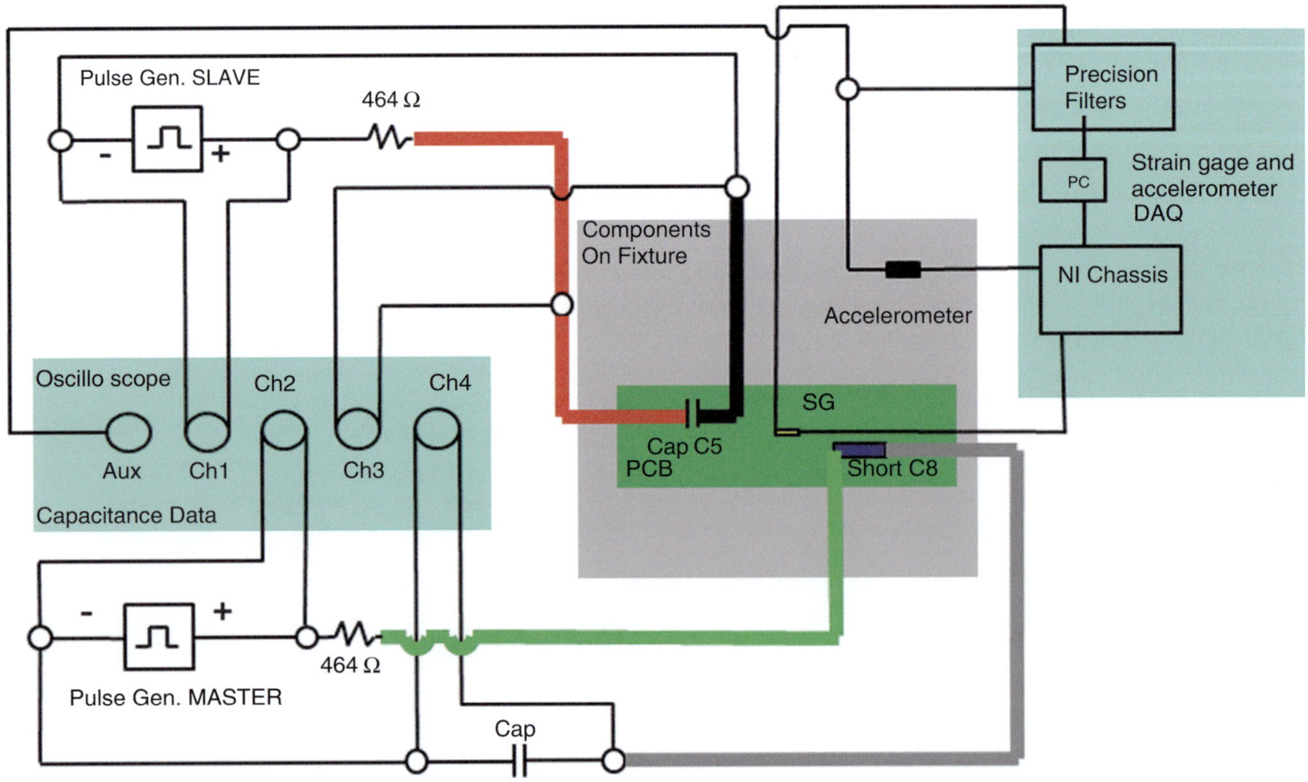

Fig. 13.3 Data acquisition schematic

constant. However, only voltages within 12 % of peak were used for curve fitting. Inductance of capacitors used in this study becomes significant at frequencies over 4 MHz, and thus was considered negligible during the dynamic capacitance measurements.

Shocked components were dismounted according to AVX corporation mounting guide instructions using 33 W MEISEI Corp HOT Tweezers with tip temperature of 210 °C as measured by an HH611A thermometer [14]. Capacitance and dissipation factor of the shocked, and control capacitors were measured at discrete frequencies of 100 Hz, 1 kHz, 10 kHz, 100 kHz and 1 MHz at +/−1 V oscillation level using an HP 4192 LF Impedance Analyzer with an HP 16048C test leads, and custom built, spring loaded test fixture. Calibration of the impedance analyzer was performed after every change in frequency for shorted and open fixture configurations.

13.2.4 Test Configurations and Matrix

Drop tests were carried out from 12 distinct drop heights using three different PCB mounting configurations for a total of 60 impacts. The drop heights ranged from 6 in. to 72 in., equivalent to 2.2 m/s and 12.8 m/s change in velocity, or peak acceleration amplitudes varying from 666 g to 23,961 g respectively on the bungee cord assisted drop tower. Printed circuit boards were assembled and affixed to the fixture using three different configurations. In Configuration A, shown in Fig. 13.4, PCB was caul clamped, allowing the board to flex after impact with capacitors mounted on stations C5, C6 and C7. Capacitance was monitored on station C5 and a short was soldered on C8 station for off-board capacitance measurements. Configuration B, shown in Fig. 13.5, utilized steel clamps and thus, dynamic capacitance was not monitored in real time, but featured capacitors on all four stations. Configuration C was an identical electrical setup as configuration A, but the mechanical mounting schemem was modified with the board being affixed directly on the flat steel surface of the fixture preventing the board from flexing post impact, as seen in Fig. 13.6. Unlike the acceleration profiles, strain data, capacitance

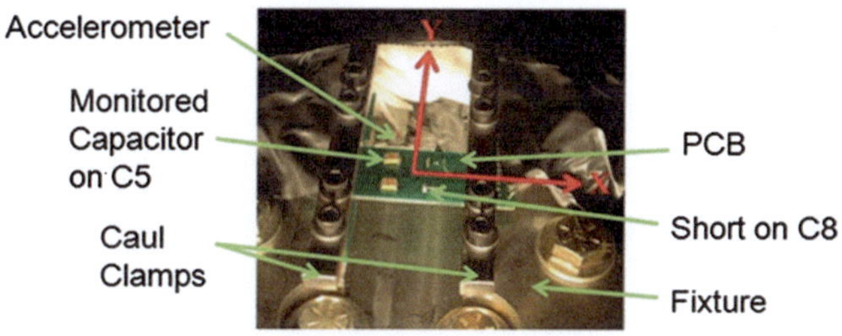

Fig. 13.4 Flexing board; Configuration A

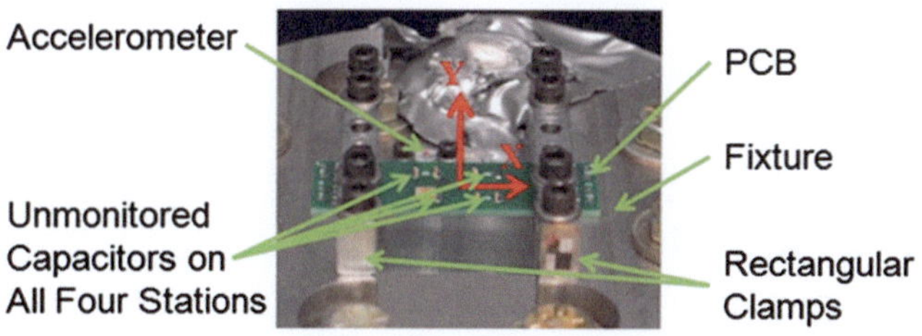

Fig. 13.5 Flexing board; Configuration B

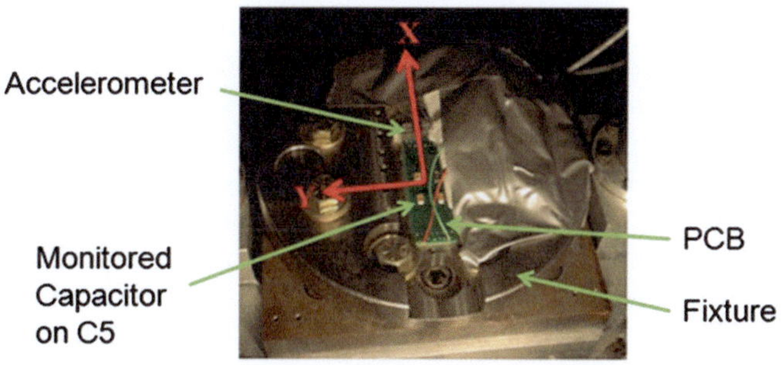

Fig. 13.6 Non-flexing board; Configuration C

Table 13.1 Drop test matrix

Test #	Config.	Height (in)	dV (m/s)	Strain data	Capacitance	Test #	Config.	Height (in)	dV (m/s)	Strain data	Capacitance
1	–	12	3.934	X	–	32	A	36	6.134	X	X
2	–	18	4.517	X	–	33	A	36	7.505	–	X
4	–	24	2.886	X	–	34	A	36	7.234	–	X
5	–	30	5.587	X	–	35	A	36	7.676	–	X
6	–	36	6.202	X	–	36	A	36	5.878	–	X
7	–	36	6.074	X	–	37	C	18	2.887	–	X
8	–	36	4.857	X	–	38	C	24	3.493	–	X
10	B	17.25	4.762	–	–	39	C	30	6.046	–	X
11	B	23.25	5.877	–	–	40	C	36	5.651	–	X
12	B	29.25	4.731	–	–	41	C	42	8.416	–	X
13	B	35.25	6.177	–	–	42	C	48	8.536	–	X
14	B	42.25	7.636	–	–	43	C	48	7.247	–	X
15	A	18	3.336	–	X	44	B	36	8.638	–	–
16	A	36	6.082	–	X	45	B	30	0.000	–	–
17	A	36	0.000	–	X	46	B	24	5.499	–	–
18	A	36	6.271	–	X	47	B	18	4.468	–	–
19	A	36	6.803	–	X	48	B	12	2.699	–	–
20	A	36	5.966	–	X	49	B	6	2.162	–	–
21	A	36	6.307	–	X	50	B	6	2.182	–	–
22	A	36	6.382	–	X	51	B	12	3.716	–	–
23	A	36	6.942	–	X	52	B	18	5.204	–	–
24	A	36	6.382	–	X	53	B	24	6.180	–	–
25	A	36	7.758	–	–	54	B	30	5.998	–	–
26	A	36	7.955	–	–	55	B	36	7.429	–	–
27	A	12	2.307	X	X	56	C	54	10.027	–	X
28	A	36	6.782	X	X	57	C	60	11.392	–	X
29	A	36	5.847	X	X	58	C	66	11.283	–	X
30	A	36	5.912	X	X	59	C	72	12.850	–	X
31	A	36	7.404	X	X	60	C	72	12.797	–	X

measurements and high speed video were not all collected during each drop, therefore, test matrix in Table 13.1 correlates the drop configuration, drop height and the corresponding change in velocity, as integrated using the trapezoidal rule, with the collected information. During drops 3 and 9, acceleration data was not recorded, and thus these tests are not included in the matrix.

13.3 Results and Discussion

13.3.1 Acceleration Profile Characterization

During experiments and with each successive drop, the felt programmer, which shapes the acceleration profile, became more impacted. As the programmer became more impacted acceleration impulse exhibited higher peak values over a shorter duration, for example—peak acceleration changed by half an order of magnitude throughout the test series where a 36 in. height drop peak acceleration increased from 6666 g to 12,340 g over the course of 60 drops, as can be seen in Fig. 13.7. However, the total change in velocity as integrated over the duration of each impulse was relatively constant as shown in Table 13.2, making it necessary to calculate this quantity in order to describe and quantify the mechanical shock for every drop. Regardless of the actual acceleration profile, strain response of the board's surface correlated well to the mechanical shock amplitude, where an increase in peak acceleration increased the peak strain amplitude and also decreased the strain frequency response, as shown in Fig. 13.8.

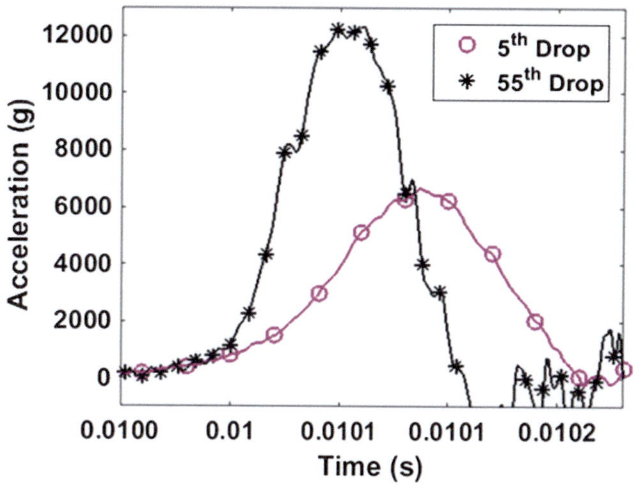

Fig. 13.7 Impacted programmer effect on 36 in. drop acceleration profiles

Table 13.2 Impacted programmer effect on peak acceleration

	Drops 1–5		Drops 51–55	
Drop height (in)	Peak acceleration (g)	dV (m/s)	Peak acceleration (g)	dV (m/s)
12	662	3.93	3217	3.72
18	1584	4.52	5526	5.20
24	2659	5.77	7940	6.18
30	4386	5.59	10,576	6.00
36	6666	6.20	12,340	7.43

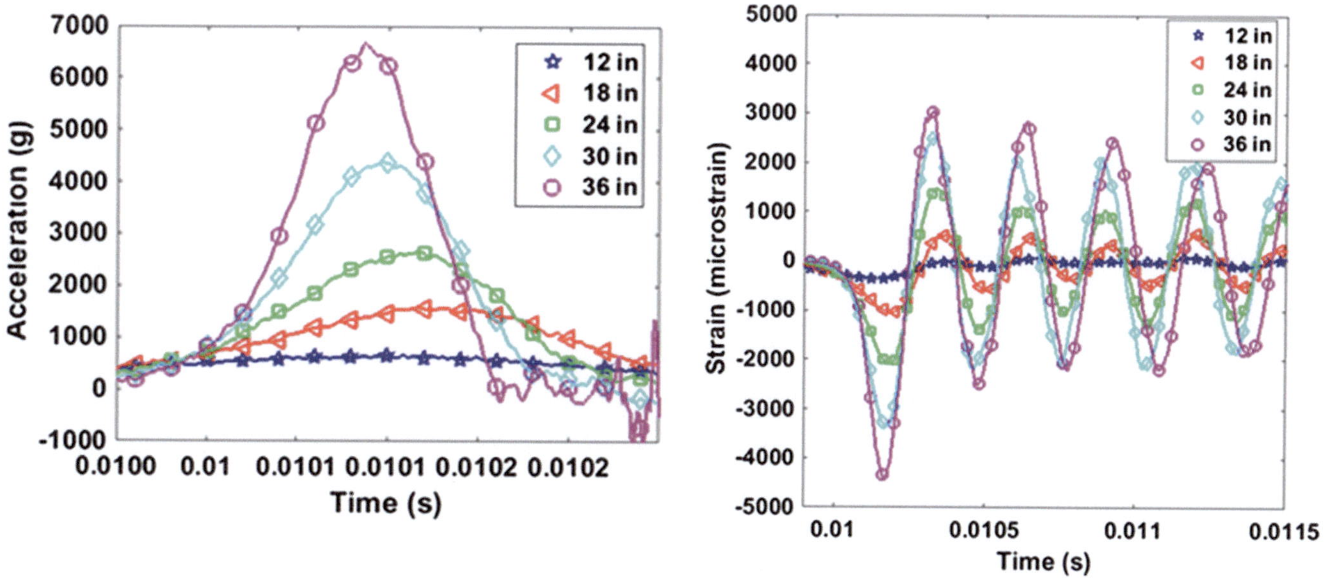

Fig. 13.8 Drop tower acceleration profiles (*Left*). Corresponding strain profiles of the top surface of PCB (*Right*)

13.3.2 Dynamic Capacitance Measurements

Dynamic capacitance calculations using least squares fit of the voltage across the stressed capacitor to the exponential functions in Eqs. (13.2 and 13.3), an example of which is shown in Fig. 13.9, did not detect any piezoelectric response of the dielectric material and thus, did not show any significant change in capacitance due to piezoelectric effect during the impact.

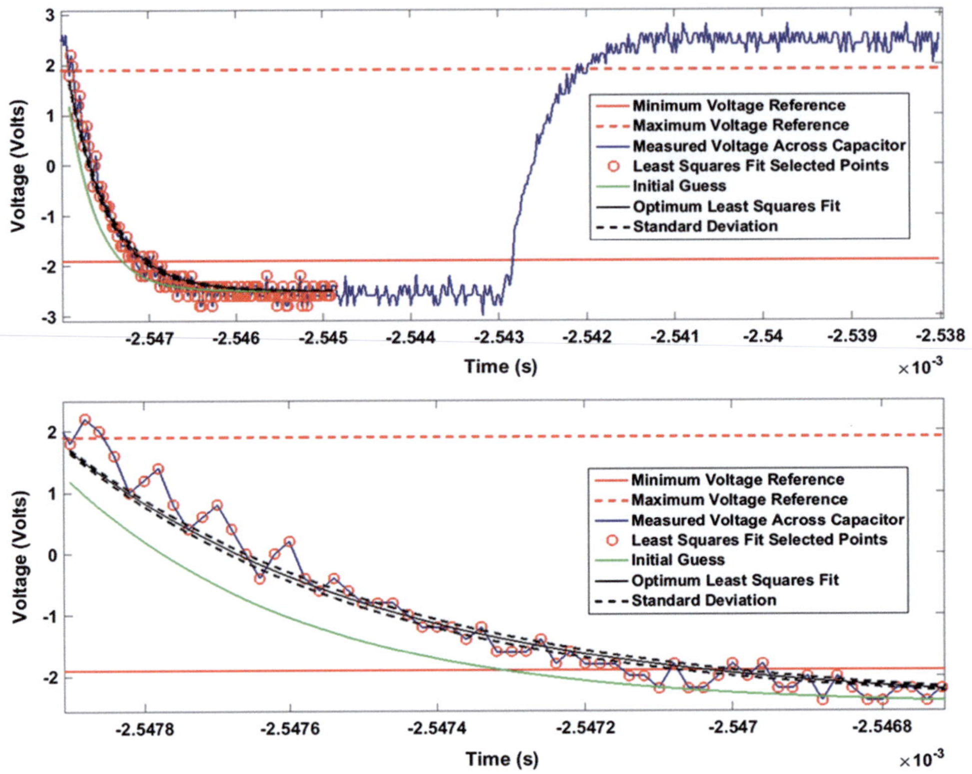

Fig. 13.9 Discharge/charge cycle at 100 kHz (*Top*). Close-up of the discharge least squares fit (*Bottom*)

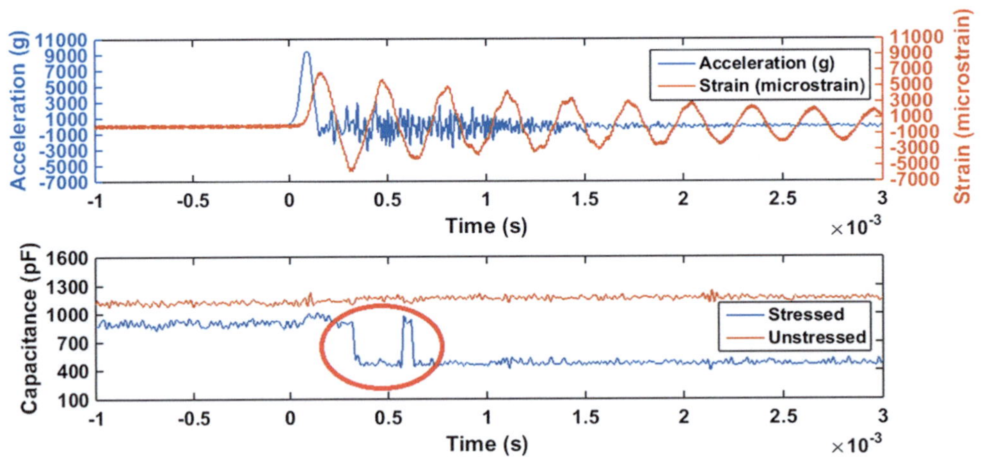

Fig. 13.10 Drop 29 Intermittent contact and final component separation

Even after the board in Configuration C has undergone a total of 100.6 m/s change in velocity over 24 separate drops with peak accelerations up to 23,961 g, no change in capacitance was detected using the RC method during the impact or when comparing the first and last drop of the board. RC dynamic technique measured and calculated values of capacitance with standard deviation of approximately 29 pF, which is 6 % of the nominal 470 pF capacitance, possibly masking any small piezoelectric effects during the impact.

However, the RC, measurement method easily detected the exact instance of the flexural failure of capacitors dropped in Configuration A. The loss of electrical contact during the flexure of the board after the impact, produced a binary capacitance profile indicative of component separating from the PCB, as can be seen in Fig. 13.10. The measured capacitance is greater than 470 pF, because 166 in. of shielded cable and the PCB itself contributed approximately 499.3 pF to the total

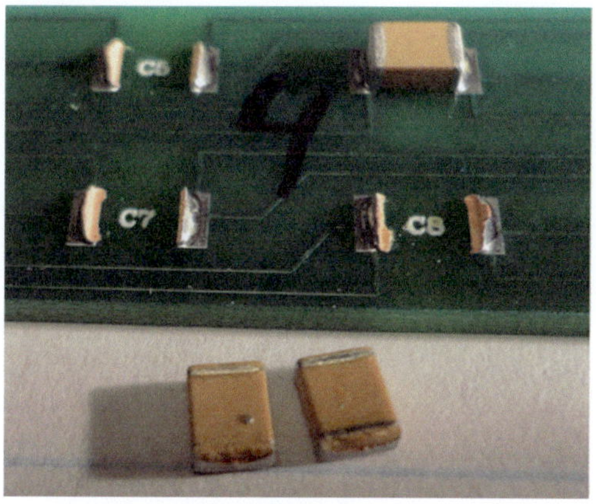

Fig. 13.11 Dominant flexural mode of failure

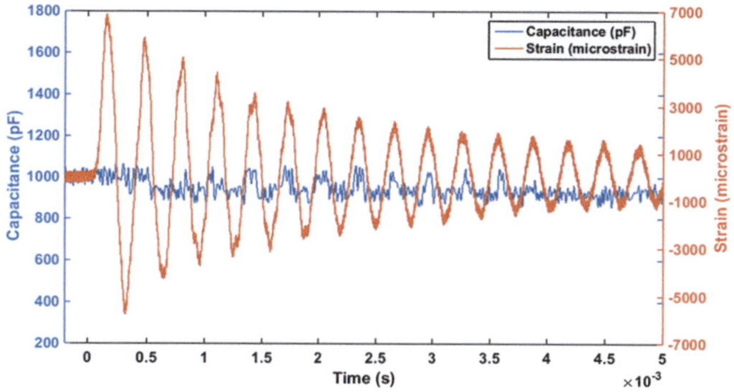

Fig. 13.12 Capacitance fluctuations correlated to PCB strain; (Acceleration and strain profiles from drop 29 and capacitance measurements from drop 28)

capacitance. Intermittent capacitance measured using the RC method, correlated well with high speed video, where monitored, stressed, capacitor dropped in Configuration A with peak acceleration of 9363 g and with a change in velocity of 5.847 m/s, was observed to first separate from the inner PCB pad, then momentarily contact the pad, before completely separating from the board's surface while the PCB was oscillating post impact. Out of 93 solder connections on the boards that were dropped from 36 in., failure rate was near 38 %. A vast majority of component failures, 91 %, were due to the flexural mode of failure, where during board oscillations, dielectric ceramic and the terminal metal were fractured, as shown in Fig. 13.11. The remaining 9 % of connection failures were due to pad cratering, but only after the opposing side of the component first separated via the flexural failure mode. With 62 % connection survival rate, flexing board performed much worse than the non-flexing board, which had no failed capacitors as measured using the RC method, under multiple drops and harsher mechanical shocks.

Before complete flexural failure, in some cases, capacitance measured with RC method exhibited an oscillatory nature at a frequency, which closely compared to the board strain oscillation response. Capacitance of the stressed capacitor and the board strain oscillated at a dominant frequency of 4651 Hz. However, not every drop produced oscillations in capacitance, nor have capacitance oscillations begun immediately after the mechanical shock. The fact that capacitance oscillations were delayed and the fact that not all shocks with similar magnitudes produced capacitance oscillations can be explained by the mechanical mode of failure of series capacitors. In series capacitors, partial mechanical failure of the floating electrodes results in proportional loss in capacitance, which can return to original values as the board oscillates and intermittently re-establishes contact between separated and failed materials, as can be observed in Fig. 13.12. Even though no strain data for the board was available for drop test 28, dynamic strain profile from drop 29 was selected to show correlation between capacitance fluctuations and board oscillations in Fig. 13.12, because both tests were conducted successively from the same drop height of 36 in. After the board oscillations are dampened by approximately 74 % of the peak surface strain, capacitance of the device is reduced by 107 pF, which is approximately a quarter of the original capacitance of the component. The 23 % reduction in capacitance is indicative of the mechanical flexural failure of a four-electrode series capacitor to an open circuit. The strain gage was mounted on the bottom of the printed circuit board for all the populated PCB drop tests, indicating that capacitance returns to the original value, when the top surface populated with capacitors is in the state of compression, as seen in Fig. 13.12.

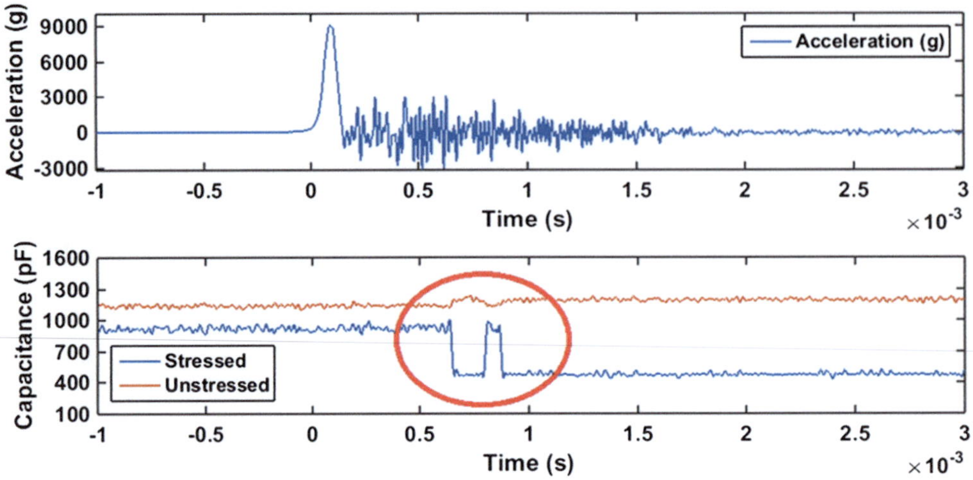

Fig. 13.13 Electromagnetic effects on capacitance measurements (drop 21)

Fig. 13.14 Effect of reflow soldering and dismounting on capacitance

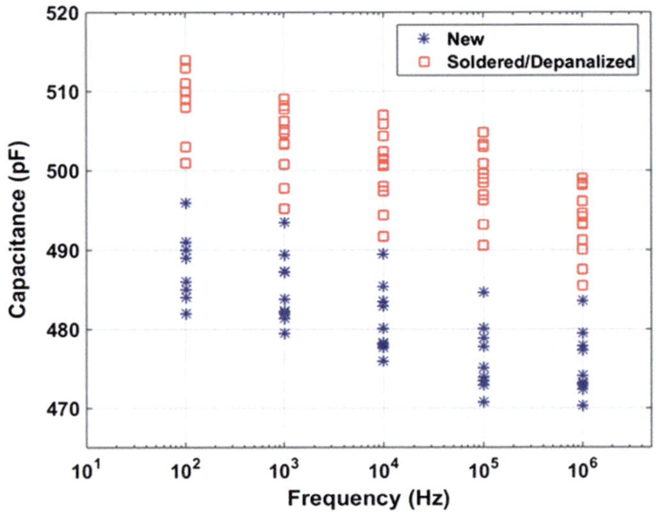

Finally, during dynamic RC capacitance measurements, electromagnetic effects where observed, where capacitance of the off-board capacitor varied with the stressed capacitor contact, as can be seen in Fig. 13.13. Capacitance value of the unstressed capacitor is increased when stressed capacitor circuit opens, and decreased again when it is closed, allowing current to flow. The variation of unstressed component capacitance is evidence of Biot-Savart Law and is due to the use of the same shielded cable for supplying the square pulse for both stressed and unstressed capacitors.

13.3.3 Static Capacitance Measurements with Impedance Analyzer

Static capacitance measurements were also conducted using an impedance analyzer on all recovered shocked, as well as on un-shocked soldered/dismounted and brand new capacitors. As with any ferroelectric material, exposing capacitor dielectric to temperatures above the Currie point, which is 120 °C for Barium Titanate, randomizes the polarization domains and increases the dielectric constant temporarily, thus increasing capacitance. Therefore, capacitance of ten brand new devices was measured and the effect of reflow soldering at a temperature of 235 °C and dismounting 12 capacitors without shocking them was determined. Across the entire frequency spectrum, from 100 Hz to 1 MHz, simply soldering and de-penalizing components increased their capacitance on average by 20.3 pF, or 4.3 %, as can be seen in Fig. 13.14. Also, as shown in Fig. 13.14, at higher frequencies a capacitor functions increasingly more like an inductor, predictably reducing its capacitance with increase in frequency.

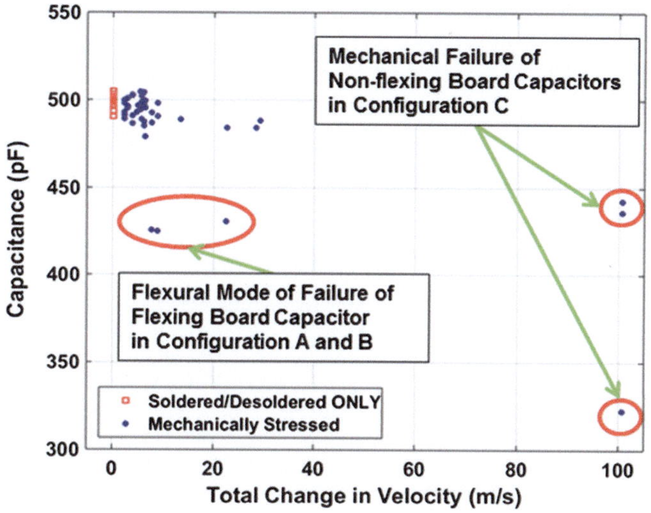

Fig. 13.15 Capacitance measured using impedance analyzer at 100 kHz vs. total change in velocity imposed on a component

Similarly to the dynamic RC method, static capacitance measurements of stressed components did not detect a significant permanent electromechanical response of the piezoelectric dielectric material for boards dropped in Configuration A or B. No clear relationship between the peak acceleration, drop height or total change in velocity and change in capacitance was evident at least for the current levels of mechanical shock, as seen in Fig. 13.15. Unfortunately, devices that showed proportional flexural mode of failure using the RC method for drops in Configuration A, separated from the PCB during the impact and were not recovered for static measurements. Abrupt changes in capacitance of devices mounted on the non-flexing board in Configuration C were possibly due to an introduction of small fractures in capacitor materials during the shock, which did not affect the capacitance of the devices significantly enough to be detectable by the dynamic RC method. However, fractures most likely grew during the dismounting process of the board and capacitors, resulting in a proportional mechanical mode of failure indicative of floating electrode capacitors, and detectable by the static capacitance measurements.

13.4 Conclusions

A novel capacitance measurement technique was developed by applying the least squares fit to an exponential function of the collected voltage across charging/discharging capacitor during the dynamic impact event. The RC dynamic measurement method was successful in identifying the exact moment of component separation, as well as correlating fluctuations of capacitance due to partial flexural failure of floating electrode capacitors and the printed circuit board surface strain oscillations after the impact. However, the RC capacitance measurement technique was accurate to only within 6 % of the nominal capacitance of the components and thus, was not feasible for detecting piezoelectric effect during the impact. Therefore, experiments suggest that even in high stress environments with high-g acceleration profiles, up to 24,000 g, mechanical, but not electrical modes of failure are still the most significant contributors to circuit breakdown in both flexing and non-flexing boards containing surface mounted components.

Also, surface mounted components on the non-flexing board were able to survive multiple impacts with more than twice the peak acceleration and a full order of magnitude higher total change in velocity than the maximum survivable conditions for the flexing board containing the same capacitor layout. Higher survivability of capacitors mounted on the non-flexing board point to a clear correlation between board dynamics and component dominant modes of failure in high-g stress environments. The relationship between board dynamics and failure modes underscores the need for PCB reinforcement in order to manufacture more reliable electronics packages, and the requirement for much closer investigation of the dynamic board response for in-depth understanding of circuit reliability. Finally, while electromechanical response of the increasingly prolific capacitor ferroelectric dielectrics was not detected at the currently tested acceleration and stress levels using the RC method developed in this work, further, and possibly computational study is warranted to determine at least an order of magnitude of mechanical stress at which piezoelectric effect becomes significant for electrical circuit stability that uses such components.

Acknowledgements The authors would like to thank the technical teams at the Air Force Research Laboratory including Alain Beliveau, Roy Kimbrough, Robert Harlow, Lazaro Sanchez, Joel Metz, James Hardin, James Hierholzer, Michael Craft and many more for their outstanding support during the experimental portion of this research.

References

1. Mechanical Shock Component and Subassembly, JESD22-B110B Standard, (July 2013)
2. Luan, J.-E., Tee, T.Y., Pek, E., Lim, C.T., Zhong, Z., Zong, J.: Advanced numerical and experimental techniques for analysis of dynamic responses and solder joint reliability during drop impact. IEEE Trans Component Pack. Manuf. Technol. **29**(3), 449–456 (2006)
3. Tee, T.Y., Luan, J-E., Pek, E., Lim, C.T., Zhong, Z.: Novel numerical and experimental analysis of dynamic responses under board level drop test. In: 5th Int. Conf on Thermal and Mechanical Simulation and Experiments in Micro-electronics and Micro Systems., pp 133–140, (2004)
4. Chou, C.-Y., Hung, T.-Y., Yang, S.-Y., Yew, M.-C., Yang, W.-K., Chiang, K.-N.: Solder joint and trace line failure simulation and experimental validation of fan-out type wafer level packaging subjected to drop impact. Microelectron. Reliab. **48**, 1149–1154 (2008)
5. Agrawal, A., Levo, T., Pitarresi, J., Roggeman, B.: Board level energy correlation and interconnect reliability modeling under drop impact. In: Electron. Comp. and Tech. Conf., pp 1694–1702, (2009)
6. Moulson, A.J., Herbert, J.M.: Electroceramics: materials, properties, applications. John Wiley & Sons, Chichester (2003)
7. Jaffe, B., Cook Jr., W.R., Jaffe, H.: Piezoelectric ceramics. Academic, London (1971)
8. Keimasi, M., Azarian, M.H.: Flex cracking of multilayer ceramic capacitors pb-free and Tin–lead solders. IEEE Trans. Dev. Mater. Reliab. **8**(1), (March 2008)
9. Prymak, J.D., Bergenthal, J.: Capacitance monitoring while flex testing. IEEE Trans. Component Pack. Manuf. Technol.—Part A, **18**(1), (March 1995)
10. Sloka, B.: Flex crack mitigation. KEMET Electronics Corporation, Tech Topics, (April 2008)
11. Xu, C-N. Akiyama, M., Nonaka, K., Watanabe, T.: Electrical power generation characteristics of PZT piezoelectric ceramics. IEEE Trans. Ultrasonics, Ferroelectrics Freq. Cont. **45**(4), pp 1065–1070, (1998)
12. Keawboonchuay, C.: Electrical power generation characteristics of piezoelectric generator under quasi-static and dynamic stress conditions. IEEE Trans Ultrasonics, Ferroelectrics Freq. Cont. **50**(10), 1377–1382 (2003)
13. Tipler, P.A.: Physics for scientists and engineers. Electricity and Magnetism, Light, vol. 2. W.H. Freeman/Worth, New York (1999)
14. AVX Surface Mount Ceramic Capacitor Products, Version 14.9, Surface Mounting Guide, pp. 117–121

MIX
Papier aus verantwortungsvollen Quellen
Paper from responsible sources
FSC® C105338

If you have any concerns about our products,
you can contact us on
ProductSafety@springernature.com

In case Publisher is established outside the EU,
the EU authorized representative is:
Springer Nature Customer Service Center GmbH
Europaplatz 3, 69115 Heidelberg, Germany

Printed by Libri Plureos GmbH
in Hamburg, Germany